How Come?

Hydronic Heating Questions
We've Been Asking
For More Than 100 Years
(with straight answers!)

Published by
HeatingHelp.com
63 North Oakdale Avenue,
Bethpage, NY 11714
Telephone: 1-800-853-8882
Fax: 1-888-486-9637
Very cool website: www.HeatingHelp.com

Printed by Searles Graphics, Inc., Yaphank, NY 11980

Manufactured in U.S.A.

First Printing, May, 1995
Second Printing, September, 1996
Third Printing, December, 1998
Fourth Printing, March, 2001
Fifth Printing, September, 2002
Sixth Printing, January, 2005
Seventh Printing, February, 2008

For "Professor" Fremont Lobbestael,

who has taught me so much

about so many things

Table of Contents

PREFACE

How Come This Book?

I've always liked the way old heating books look, the way the Dead Men drew the pictures of the equipment freehand with pen and ink (such sharp lines!). The pages in those old books have the smells of a hundred years pressed into them like dried flowers. They crumble if you're not careful.

I sit, most days, and think about the people who owned these books before they came to rest on my bookshelves. Dead Men wrote notes in the margins along with those numbers, formulas, and reminders of what they needed for that next job. The notes echo in the margins like voices from the past.

And the authors speak so clearly, even now. I remember the first time I met Theodore Audel. He was dead long before I was born, but nevertheless, he walked me through the basics of hydronic heating by using that beautifully simple question-and-answer method of his. Reading his book was like studying a catechism. How come? Because... How come? Here's why... How come? Look at Figure 107... After awhile, it all made sense. It was that question-and-answer method of his.

I met Alfred G. King, another famous Dead Man, on the top shelf of a used bookstore in Lancaster, Pennsylvania. His book cost me five dollars. Mr. King explained the workings of the old hot water systems to me in a way that made it so simple. He, too, used that question-and-answer format. He knew what I needed to know, and he put the information in order for me. He spoke plain-English and he was patient. He was always willing to go over it again if I didn't get it the first time. He never yelled, and he never called me an idiot. A real gentleman, Mr. King.

R.M. Starbuck came to me through my friend Bob Steinhardt. Bob stopped by one day with a Starbuck question-and-answer book about plumbing. Not long after, I learned that R.M. Starbuck also had a lot to say about hydronic heating. I met him in another old bookstore. He'd been waiting for me.

In 1993, I met Doug Starbuck, R.M. Starbuck's descendant

and the current Caretaker of Heating History. Doug reminded me
that there's more to this trade than just making a living. He
showed me things I didn't know still existed, and he warmed me
with an enthusiasm you hardly see any more. He loves this trade,
and he has so much to say. Listen:

"One time I went out to a house they were building on the
other side of town," he said. "The carpenter was working with
this kid apprentice. I got there at about three o'clock in the afternoon and sat down on a nail keg. I sat there for about an hour.
The apprentice kept looking over at me. Finally, he asked the
carpenter what I was doing. The carpenter looked over at me and
went back to hammering. 'Doug's working,' he told the kid.
'Leave him alone.'

"The next morning I showed up with my drill and punched
out all the holes I needed to do the whole job. I put the drill
away and never took it out again. When I was piping the job out,
the carpenter said to the kid, 'See, you work first in your head,
and then you pick up your tools. That's the right way to do
things.

"If you have it in your head…you own it."

R.M. Starbuck speaks through his distant nephew, Doug. So
much of the trade works like that—from father to son to nephew
to brother to sister. And most of the information gets passed
down through word of mouth, in plain-English, by someone asking How come? and another answering Because…

So that's how come I wanted to write this book. I wanted to
pass on what I've learned so far, and I wanted to do it in the
same way the Dead Men did it. I wanted to pay them some
respect.

I've chosen the topics carefully. Gravity hot-water heat came
first, so it should rightfully be Chapter One, and it is. Within this
chapter you'll find so many of the basic principles that make all
the other systems work. Read carefully, and your days will go
smoother, no matter whose basement you're in.

Next comes Indirect Hot-Water Heating, one of the most
intriguing areas of hydronic heating. It can scare the heck out of
you the first time you bump into it, and it can have you scratch-
ing your head in wonder, but here, too, you'll find many of the
basic principles that apply today—even within our most modern
hydronic systems.

From there, we'll move into Diverter-Tee Hot-Water Heating
and see how the old-timers applied much of what they'd learned
during the Gravity Era to the homes they built in the Thirties

and Forties. There are a lot of subtleties here, and a lot of opportunities to shine when it comes time to troubleshoot.

Following World War II, Loop Hot-Water Heating became popular because it was so easy to do. We needed this system to come along at that time because of the tract housing and the building boom that propelled the trade through the Fifties and on into the Sixties. We were in a big hurry back then and as a result, many of these systems have never lived up to their promises. In this chapter, I've tried to show you ways you can make this simple system better.

Radiant-Floor Hot-Water Heating has been around for a few thousand years, but suddenly, it's brand-new again! There are a lot of questions when it comes to this subject. In Chapter Five, you'll find a long list of the questions I've asked over the years, along with the straightest answers I could possibly give you. I hope they shed some light and help you take advantage of this rapidly growing market.

Finally, I'm going to tell you about Condensate Hot-Water Heating because it combines a wonderful mix of the past and the present along with some creative solutions to tough problems.

So how come this book? Because most of the time, all we really need are straight answers to direct questions—but we're not always sure who to ask.

I hope you find what you're looking for within these pages.

CHAPTER ONE:

GRAVITY
HOT-WATER
HEATING

"The construction of the piping system affects the results with steam heating, but it has a much greater effect with hot water systems."
—Professor Rolla C. Carpenter, 1899

Q: How long has gravity hot-water heating been around?

A: Gravity hot-water heating began quietly in the United States between 1875 and 1885. It was a Canadian import, a safe substitute for steam heat, which had been earning a notorious reputation throughout the world as being a pretty dangerous way to heat a building.

Q: What was wrong with steam?

A: The trouble with steam in the early days was that it ran under pressure and frequently exploded with disastrous results. Hot-water systems, on the other hand, were open to the atmosphere, and relatively safe because the old-timers usually limited them to a high temperature of 180°F. In those days, you could liken the difference between a gravity hot-water system and a steam system to that of an open, simmering pot of water, and a pressure cooker gone berserk!

Q: So gravity hot-water became popular because it was safe?

A: Yes, and because these systems were also easy to maintain and, most of the time, operated with little or no trouble. They had a lot going for them, and they quickly became the preferred way to heat large American homes just before the turn of the century.

Q: Is it a simple system?

A: In theory, it is. The only moving part is the water itself, but to get that water to go where he wanted, a pipe fitter had to meld the knowledge and experience of Mr. Goodwrench and Mr. Wizard. If he did his job well, the system worked beautifully. If he didn't, it became a balancing nightmare.

Q: What did a typical gravity hot-water system look like?

A: Here's a diagram of an "upfeed" system.

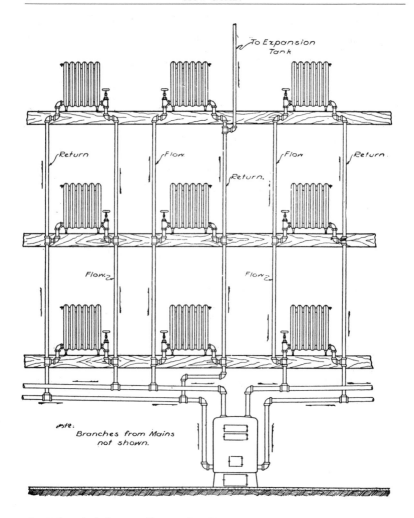

Q: Why did they call it upfeed?

A: Because the water fed up from the bottom (the boiler) to the top (the highest radiator).

Q: Where's the circulator?

A: There isn't one! Circulating pumps, which we use on modern hot-water systems, hadn't been invented yet so to move water from the boiler to the radiators, the old-timers depended on a basic law of physics: Hot water rises, cold water sinks.

Q: Why is that?

A: Because of the difference in density between hot and cold water. A cubic foot of water at 180°F takes up about five percent more space than a cubic foot of water at 40°F. It also weighs about two pounds less.

Q: Is that where the term "gravity" comes in?

A: Yes! When you heat water in a boiler, it will rise up into the pipes because it's lighter that the relatively cold water in the system piping. That colder water, in turn, falls back down into the boiler (by gravity), and before long, you have a Ferris wheel flow of warm water moving freely from the boiler to the radiators.

Q: What determines how fast the water moves?

A: Several things. First, there's the height of the system. The taller the building, the quicker the flow. Within reason, of course, because if the building is too tall, the water will cool and slow circulation to the upper floors. A three-story house is the practical limit for gravity hot-water heating.

And then there's the size of the pipes. The larger the pipes, the faster the water will flow. This is because large pipes offer less resistance to flow than small pipes. It's also the reason why the old-timers used two supply and two return tappings on their boilers.

Ultimately, the size of the pipes was also the reason steam replaced gravity hot-water heat in American homes. As the years went by, steam heat became safer, but the large-diameter pipes the gravity systems required continued to be expensive.

The third factor that determines how quickly the water circulates is the condition of the pipes. When the pipes are new, they're smooth on the inside. They offer very little resistance to the slow-moving water. However, as they age, the pipes develop little nooks and crannies because of oxygen corrosion. These tiny internal burrs increase frictional resistance, and that, in turn, slows the flow and the movement of heat to the radiators. Nowadays, we usually overcome this problem by adding a circulator to the system.

Finally, there's the difference in temperature between the supply and return water. The hotter the water, the faster it circulates. However, the old-timers always kept the maximum temperature at 180°F to make sure the water never approached the boiling point.

Q: Did the old-timers work with a certain temperature difference between supply and return?

A: Yes, and to get the best efficiency, they limited the maximum temperature difference between supply and return to 20°F. This was a function of pipe sizing (the smaller the pipes, the greater the temperature drop, and vice versa). So on the coldest day of the year, if water left the boiler at a maximum of 180°F, it would return at a minimum of 160°F. This assumes, of course, that the pipe fitter followed the accepted piping practices of the day.

Q: Did the hot water take up more space than the cold water?

A: It sure did! As I said before, when you heat water from 40° to 180°F, you wind up with about five percent more water than you started with. You have to have a place to put that "extra" water.

Q: How did they deal with the "extra" water?

A: They used expansion tanks.

Q: What does an expansion tank look like?

A: A typical one looked like this.

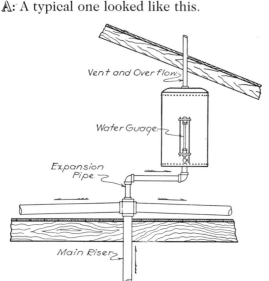

Vent and Overflow

Water Guage

Expansion Pipe

Main Riser

Q: Where did the expansion tank go?

A: Typically, at the high point of the system. You'll usually find them in the attic. The tank gives the expanding and contracting water a place to rise and fall.

Q: Suppose I put too much water into the system

when I first fill it up. What will happen?

A: It will overflow from the tank through its vent and wind up on the roof.

Q: **Can this do any harm?**

A: Not to the system. It might leave some rust stains on the roof if the system is old, but that's about it.

Q: **How much water should I put into the tank when I'm first filling the system?**

A: Normally, you should maintain the tank at one-third full when the water is cold (there's often a gauge glass on the side of the tank so you can see what you're doing). As the water heats and expands, it will rise into the upper two-thirds of the tank and stop before spilling over onto the roof.

Q: **How did they fill these tanks?**

A: Some tanks had an automatic fill valve, which is very similar to the ballcock in a toilet tank. Others, the old-timers filled by hand with a valve that was either down in the basement or up in the attic.

Q: **Wait a minute, if you're in the basement, how can you tell how much water is in the attic tank?**

A: Good question! Chances are the boiler had an "altitude" gauge that showed the height of the water in the system. The gauge registered feet of altitude as well as static pressure.

Q: **What's static pressure?**

A: It's the pressure created inside the boiler by the water as it stacks up in the system piping. The gauge records static pressure in "pounds per square inch" (psi). One psi will lift water 2.31 feet (that's 28 inches) straight up, and that's where the term "altitude" comes in.

Q: **Do you have to take any special precautions with the upfeed gravity system?**

A: Yes, if you have to drain the system, be careful how you refill it. Start off with all the radiator vents open. Then, slowly fill the system, one floor at a time. When water flows from the vents on the first floor, quickly close them all. Then, continue filling until water rises to the second floor. Shut all the air vents and move up to the third floor. Once you have all the radiators filled, fill the system to the one-third-full point in the expansion tank.

Q: Why is this method important?

A: Because there's so much air in those large pipes and radiators. If you try to fill the system all at once, and then go back and bleed each radiator, the escaping air from one radiator will cause the water to drop out of the expansion tank and the nearby radiators. This can pull more air into the system piping.

Q: What happens if I don't follow this fill procedure?

A: Usually, you'll wind up with "phantom" air problems. The air appears in this radiator today. You vent it. Tomorrow, it's in that radiator over there. You vent it. The next day, the problem appears somewhere else. It can be maddening.

Q: How does the air from the heated water get out of the system after the initial purge?

A: It vents out through that overflow pipe that sticks out through the roof. Usually, the tank sits atop the main system riser at a high point. The tank vents most of the air that the heated boiler water releases. Should some of this air wind up in the radiators instead of in the tank, it can slow the flow of heat to the rooms. Ideally, with this type of system, someone should bleed the radiators at the start of each heating season.

Q: Is there a danger of the attic tank freezing if the attic isn't properly insulated?

A: Yes, there is. And if that happens, the expanding system water will have nowhere to go. To avoid this potentially dangerous situation, many old-timers piped their tanks like this.

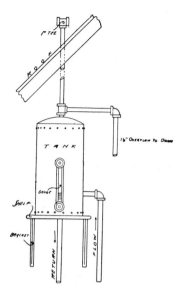

That second pipe, connected into the side of the tank, allows hot system water to circulate through the tank. Since the water is hot and in motion, it's much less likely to freeze.

Q: Why didn't they just go ahead and pipe all their tanks this way?

A: Because by circulating the water through the open tank in this way, the rate at which water will evaporate from the system increases. That means someone has to add more fresh water. Fresh water increases the rate of corrosion in the system and, over time, slows circulation.

Q: Suppose I decide to modernize the system by adding a circulator or replacing the boiler. Should I keep the open tank?

A: In a case such as this you'll probably want to close the system by replacing the open, attic expansion tank with a closed compression tank. This isn't always necessary, but it does cut down on the corrosion that takes place in the system.

Q: What's the difference between an expansion tank and a compression tank?

A: It's really just a question of semantics. An "expansion" tank is an open tank. A "compression" tank is a closed tank. Most people interchange the terms. As long as the person you're talking to knows what you mean, it really doesn't much matter what you call it.

Q: Were there other types of gravity systems?

A: Yes. If the original owner of the home went first-class, he would have installed an overhead gravity system such as this one.

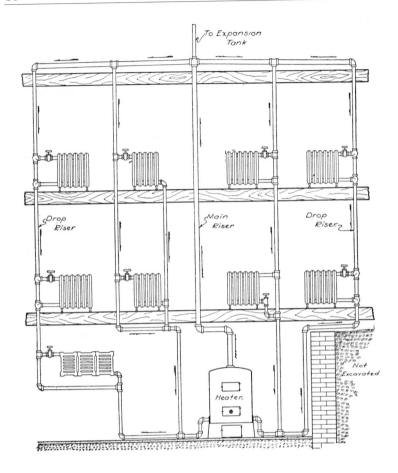

Q: How does the overhead system differ from the upfeed system?

A: In the overhead system, water goes first to the attic (or to a main suspended from the top floor ceiling) and then feeds down to the radiators. Because this "express riser" is very large, it offers less frictional resistance to the water. As a result, the hot water moves more quickly from the boiler to the radiators than it would in the upfeed system.

Another plus is the way the cooler water pulls the hot water through the radiators as it falls down the return risers. This force counteracts the effects of friction and makes the radiators heat faster. As a result, an overhead system generally costs less to operate.

Q: Is this type of system easier to vent?

A: Yes, much easier. In fact, because of the way the radiators are connected to the mains, you don't need radiator air vents with this system. All the system air vents automatically through the attic tank. It doesn't take long to fill this system either, and you don't have to worry about spilling water all over the floor while venting, as you do with the upfeed system.

Q: How did they pipe the radiators into the mains on this system?

A: They always used top and bottom connections. They could enter the top on one side of the radiator and leave through the bottom on the opposite side, or they could enter and leave through the same side. This second method saved a riser, which made for a less-expensive installation.

Q: Didn't they need special fittings to make this work?

A: Yes. They had to divert water through the radiator. To do this, they used a special type of tee. Here's a picture of one.

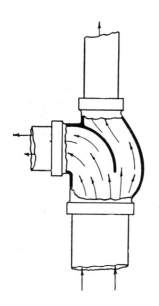

Q: What did they call this tee?

A: They called it an "O-S" fitting after its inventor, Oliver Slemmer of Cincinnati, Ohio. It was a beautifully simple device.

Q: Is this similar to a "Monoflo" tee?

A: It is, but the O-S preceded the Monoflo by many years. During the 1930s, the Bell & Gossett Company introduced their "Monoflo" tee (the name is a trademark). It went on to play a big part in American house heating during the years before World War II.

Q: Do these special tees "tell" the water where to go?

A: In a sense, they do. They create a path of least resistance for the water and direct it toward the radiator.

Q: **Is there any other way of directing the water in this type of system?**

A: There are a *number* of ways, and all of them are critical to the system's operation.

Q: **Why is this?**

A: Because the pipes in a gravity system are very large and contain a lot of cold water on start-up. Not all of that water is going to get hot at the same time. And since hot water is lighter than cold water, it has a tendency to shoot directly up to the top-floor radiators—just like a hot-air balloon. That's its path of least resistance.

Q: **So the top floors tend to heat more quickly than the bottom floors in a gravity system?**

A: Yes, and that leads to system imbalance.

Q: **How did the old-timers get around this problem?**

A: They sometimes added orifice plates to the top-floor radiator hand valves. Here's what one looks like.

Q: **What exactly is an orifice plate?**

A: It's a round piece of metal with a small hole drilled through its center. You could make one yourself out of sheet metal; most of the old-timers made their own.

Q: How did the orifice plate direct the water?

A: By increasing the resistance through the radiator it was assigned to. If water found it difficult to enter, say, a top-floor radiator because of the orifice plate, it would go to a radiator on a lower floor instead. In this sense, the orifice plate was similar to the "O-S" and "Monoflo" fitting. The big difference, however, was that instead of directing the water *into* the radiator it was assigned to, an orifice plate directed the water *away* from that radiator.

Q: What other methods did the old-timers use to make the water go where it was supposed to go?

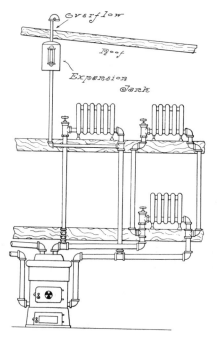

A: More often than not, they'd pipe the job in such a way as to avoid the problem in the first place. Here, take another look at this upfeed system.

We have three radiators—two on the second floor, one on the first. The hot water's tendency is to race up to the second floor. But look closely at the way the fitter makes his lateral take-offs from the supply main. Notice how the hot water supply to radiator #1 comes off the *side* of the main. The fitter did it this way because on start-up, the hottest water will be at the *top* of the supply main.

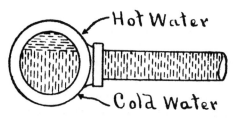

That hottest water *wants* to go to radiator #1 but it can't get there right away because the water near the bottom of the horizontal main is colder than the water near the top of the horizontal main. That colder (and heavier) water is crowding the hotter water out of the way and driving it toward radiator #3, which just happens to be on the first floor.

Q: So you can tell from the basement where the risers are going?

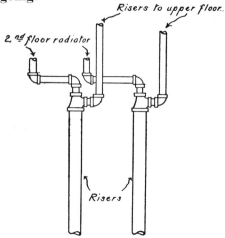

A: Yes! They usually fed upper-floor radiators from the side of the main and first floor radiators from the top. That way, the system went into more of a natural balance.

Q: Did they do similar things with their vertical risers?

A: Yes, they did. Frequently, they'd supply a second-floor radiator from the top of the riser and a third-floor radiator from the side of that same riser.

In this case, the second-floor radiator is the lower of the two. That's why it gets the water from the top of the riser.

Q: How about the horizontal mains? Did the old-timers use the same size all the way around the building?

A: Not usually. It was customary to reduce the size of the supply main as it worked its way around the building, but if the fitter reduced the pipe too quickly, flow would stop because there would be too much overall resistance.

Q: What rules did they follow?

A: Generally, they wanted the internal traverse area of the main to meet or exceed the internal traverse area of all the attached radiator hand valves. If the main was too small (or if someone added radiators to an existing main) some radiators wouldn't heat well. The competent fitters sat and calculated every job they worked on. They knew no two were quite the same.

Q: What's internal traverse area?

A: Look down the round end of the pipe. The interior circle at the open end represents the internal traverse area. By using mathematics, you can figure out how many square inches of space there is inside that circle.

Q: Can you give me some examples?

A: Sure! Here's a list of common pipe sizes used in gravity systems.

Pipe Size	Internal Traverse Area (in Square Inches)	Pipe Size	Internal Traverse Area (in Square Inches)
1"	0.86	3½"	9.89
1¼"	1.5	4"	12.73
1½"	2.04	5"	19.99
2"	3.36	6"	28.89
2½"	4.78	8"	51.15
3"	7.39		

Q: How about the supply and return mains? Do they have to be kept close together?

A: Yes. Ideally, the return main should parallel the supply main within a distance of no more than 8½ inches. It should drop only when it reaches the boiler room.

Q: How did the old-timers bring their returns from their radiators back into their mains?

A: They followed this rule: Returns from first-floor radiators have to enter on the *side* of the return main because they leave from the *top* of the supply. This is important because the return from one radiator could block the return from another if the temperatures coming back from the two radiators are slightly different, which they almost always will be.

Q: Were there any special fittings for the mains?

A: They used a number of them. Here are two examples of the

more common ones. This is called a Eureka Fitting.

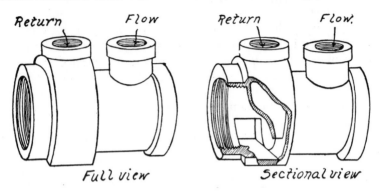

This one was known as the Phelps Single Main Tee.

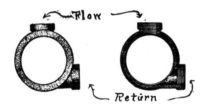

Notice how the hot water leaves from the top of the fitting while the cold water flows back into the side. Those old-timers were clever, weren't they?

Q: Is it difficult to troubleshoot gravity hot-water systems?

A: Troubleshooting *can* be tricky. There may be places in the system where hot and cold water pass each other in the same pipe. That might be perfectly normal, but you'll need to "see" it in your imagination to figure out what's going on.

Some problems may have existed for years before you got involved. Something as simple as an unreamed pipe can stop a radiator from heating, but then, so can the corrosion that builds up after sixty or seventy years of operation. You'll have to think clearly, and ask lots of questions.

Q: Does the water flow the same way here as it does in a forced circulation system?

A: Not at all! In fact, gravity hot-water heat is the mirror image of forced hot-water heat. When you use a circulator in any system, the path of least resistance will *always* be the shortest (lowest pressure drop) loop because that's the path with the least resistance to flow. Water is lazy, and when you pump it, it always wants to get back to the pump's suction as quickly as possible. Remember, in a gravity hot-water system, the path of least resistance is the top floor, which is usually the *longest* run. That's the opposite, the mirror image, of a pumped system.

Q: Can you give me a visual example of the difference?

A: Well, when I'm troubleshooting gravity hot-water heat I always think about the convective currents in a heated room. Here, think with me.

Air rises out of a radiator because it's hot and light (the same reason water rises out of a boiler). The air creeps across the ceiling and gives up its heat to the things it touches (as the water gives up its heat to the radiators). As it cools, the air in the room gets heavier and falls (the same way the water falls out of the radiators). Finally, when it reaches ground level, the now relatively cold air (like the relatively cold water inside the gravity system) scoots across the floor (or, in the case of the water, back toward the boiler) and enters the bottom of the radiator to replace the rising hot air.

But now, suppose you turned on a ceiling fan in that heated room. You'd change that convection current in a hurry, wouldn't you? You'd be "pumping" the air around the room instead of letting it rise and fall by its own buoyancy. It would go where the resistance was least when the fan was on, wouldn't it? Sure it would—just as hot water moves where the pump tells it to go.

That's the difference between gravity hot-water heat and forced hot-water heat. One moves by natural convection, the other at the will of the pump.

Q: Can those orifice plates we looked at before cause system problems?

A: Sometimes. When you add a circulator to a gravity system, the path of least resistance naturally shifts to the first-floor radiators because that's the shortest path back to the boiler. The water doesn't want to go to the top floor anymore. Those orifice plates are in the top-floor radiators. The old-timer put them there to force water down to the lower floors.

Q: What's wrong with that?

A: Well, now that you're pumping the system, the orifices are going to make *sure* the resistance through the top-floor radiators is always greater than it is through the lower floor radiators. In fact, once you add the circulator, you'll probably have no flow at all through the top-floor radiators!

Q: Won't you be able to tell right away that the orifices are causing the problem?

A: Probably not, because this problem looks exactly like an air problem. Think about it. The trouble is on the top floor. You may have drained the system when you installed the circulator. And now the folks have no heat. It looks like an air problem, but it's really a flow problem.

Q: How will I know it's a flow problem?

A: When you bleed the radiator you won't get any air. And if you don't get any air, it ain't an air problem!

Q: So what's the solution?

A: Take the orifice plates out of the top-floor radiators and stick them into the first-floor radiators. In other words, reverse the mirror image. The system will go into balance and that phantom "air" problem will be just a bad memory.

Q: Is there anything else I need to watch out for?

A: Yes, painters! If you have a sudden no-heat call on the lower floor of a gravity hot-water system, check to see if someone recently removed the radiator to strip the paint from it (or removed the radiator to paint the wall behind it). Painters and paint-strippers often close the hand valves and disconnect the radiators to make their job easier. When they do, the orifices usually fall out of the hand-valve unions. Since the average painter doesn't know beans about heating (gravity or otherwise) he doesn't know what to do with the orifice plate. To him, it looks like a piece of junk. He'll throw it in the garbage, and figure he's doing the owner a favor by "getting rid of that lose hunk of metal that was clogging up the pipes and blocking the heat." Without the orifice plate, however, most of the water will flow to the top floor.

Q: When is it a good time to convert a gravity hot-water system to forced circulation?

A: Usually, when the gravity system slows down because of the corrosion that has taken place over the years. Those little nooks and crannies in the pipe slow the flow and stop the heat. The natural response is to raise the temperature to make the water circulate more quickly. But you can only push the temperature so far before you begin to ask for trouble. That's when it's time to convert the system to forced circulation.

Q: What does this involve?

A: You have to add a circulator and (usually) close the system to atmosphere. You'll also have to make some changes to the near-boiler piping.

Q: What changes?

A: The old boiler probably has two outlets and two inlets because the idea in those days was to get the greatest possible gravity-induced flow of water through the boiler. The more holes, the better the circulation. That piping looked like this.

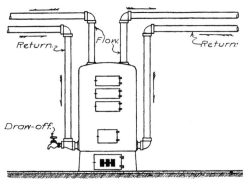

When you add the new circulator, you won't need to use such big pipes coming and going out of the boiler. In fact, you'll want to reduce the size of your near-boiler piping to give the circulator something to "push" against.

Q: Why does the circulator need something to "push" against?

A: So it won't kick itself off on its internal overload protector. A circulator does its maximum work when there's little or no resistance to flow. In a gravity system, the large pipes can't offer much resistance.

Q: **Will I still need those double inlets and outlets at the boiler?**

A: No, and that's another reason you should rework the near-boiler piping. With two inlets and two outlets, the pumped flow might short-circuit around the boiler without moving out into the system.

Q: **Suppose I don't want to repipe the boiler?**

A: You may have to use two circulators—one on each supply line.

Q: **How will I know what size pipe to use on the new boiler?**

A: A good rule of thumb is to take the largest pipe, divide it in half and then drop one size from that. That becomes the size of your new near-boiler piping. For instance, let's say the largest pipe is 2½" (if there are two inlets and outlets, you only have to consider one of them). Divide that in half to get 1¼". Now drop down one size to 1" and that's what you'll use all around your new boiler.

If your largest size happens to be two-inch, pipe your new boiler in ¾". It will look odd, and it might make you feel uncomfortable, but it'll work. Different systems call for different piping techniques. One size *doesn't* fit all and a gravity conversion is definitely different from a brand-new, forced-circulation job.

Q: **How do I size the circulator for a conversion job?**

A: It's real easy with these jobs. You're looking for high flow at a relatively low head pressure. A good choice is a circulator similar to Bell & Gossett's Series 100.

Your goal is to move a lot of water around the system as quickly as possible against very little resistance to flow. This type of circulator does just that.

Q: **Can't I use a small, water-lubricated circulator instead?**

A: These are fine circulators for most modern, forced-circulation systems, but not the best choice here. You don't need to generate much head pressure on these conversion jobs because the pipes are enormous and the resistance to flow is almost nonexistent. Using a small, high-speed, wet-rotor circulator is a poor choice on a gravity conversion because it will do the exact opposite of what you're trying to accomplish.

Q: I'm not sure I understand the difference between flow and head pressure. Can you explain it?

A: Sure! Flow is the "train" on which heat travels. Flow "delivers the goods" to the radiators. Head is *resistance* to flow and it's important, too, but only in relation to flow.

Q: Well, then what determines the head pressure?

A: In general, the size of the pipes. The smaller the pipes, the greater the required pump head, and vice versa. Since gravity systems have very large pipes, there's no need for a high-head circulator. What you need is high *flow*.

Q: Where is the best place to install the circulator?

A: It's always best to put it on the supply side of the boiler, pumping away from the compression tank. Piped this way, the circulator will add its pressure to the system's fill pressure and make it easier to get the air out. The system will also run more quietly.

Q: Do I have to use a bypass around the boiler on these jobs?

A: Most boiler manufacturers recommend that you install a bypass around their new boilers when you use them on a gravity system. Here's what that bypass piping looks like.

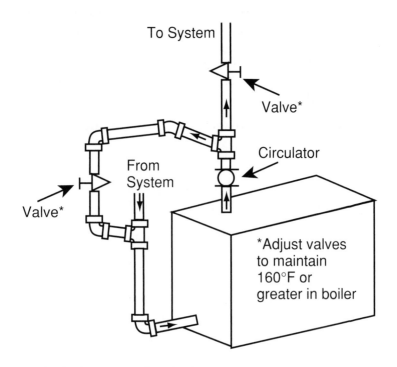

Q: What's the reason for the bypass?

A: It's there to protect the boiler against condensation and thermal shock.

Q: What's thermal shock?

A: Thermal shock is what happens to hot metal when you hit it with relatively cold return water. If you take a glass plate out of the oven and run cold water over it, it will break, won't it? That's thermal shock.

Q: How does the bypass piping help prevent this?

A: The boiler bypass allows the majority of the supply water to bypass *around* the boiler while just a small portion of the return water flows *through* the boiler, picking up the necessary heat.

Q: You said something about condensation. What's that all about?

A: If the return water temperature is too cool, the combustion gases can reach their dew point and turn into a liquid inside the boiler. That liquid is very corrosive to metal. It can damage or destroy a boiler in no time at all. By using the bypass, you're mixing hot supply water into the relatively cold return water and raising the boiler water temperature to a point where the gases can't condense inside the boiler.

Q: Does the bypass serve any other purpose?

A: It allows the boiler to come up to high-limit temperature and shut off. Without the bypass, the large volume of water moving through the boiler often keeps the temperature low and prevents the boiler from reaching high-limit. This does a good job of increasing the fuel bill.

Q: Is there another way to pipe the replacement boiler without using the bypass?

A: You can use primary/secondary pumping techniques.

Q: What's primary/secondary pumping?

A: It's a way of treating the flow through the system and the flow through the boiler as two separate things.

Q: Is there an advantage to this?

A: There is because some boilers require a minimum flow to operate at their maximum potential. This flow may not be the same as the flow you need in the system. By using primary/secondary, you're giving the system what it needs while protecting the boiler from thermal shock and condensation.

Q: How do I pipe for primary/secondary flow?

A: Tie the existing supply and return lines together to form a system loop. Then, use two standard tees, set no further than a foot apart, and attach the new boiler to the loop. Like this.

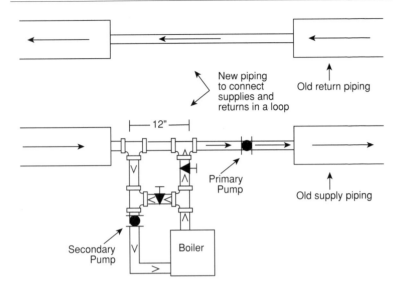

The primary pump serves the system, while the secondary pump takes care of the boiler. You meet the flow needs of both in a very simple way. The not-more-than-twelve-inch spacing between the tees allows the pumps to operate independently. When the secondary pump is off, there will be no flow through the boiler if you keep the spacing within that 12" limit.

Q: Why is that important?

A: By controlling the flow through the boiler, you're taking charge of the standby losses of the system. If the burner is off and the boiler pump is stopped, you will have minimal loss to the flue.

Q: How do I control a primary/secondary system such as this?

A: You can have both pumps and the burner come on at the same time. Or better yet, you can run the system pump (the primary) on an outdoor-air reset control, and cycle the boiler pump (the secondary) and the burner to meet the temperature needs of the building on any given day. This is the ideal way to manage an old gravity hot-water system.

Q: Can I use more than one boiler with this type of system?

A: You sure can! This system is ideally suited to a multiple-boiler setup. Watch.

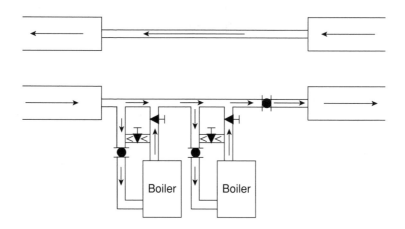

Here, we're using two boilers instead of one. The primary pump moves the water through the radiators. The secondary (boiler) pumps come on to bring a portion of the primary flow through the boilers. On mild days, you'll use just one boiler; on colder days, the boilers will piggyback to bring the water temperature up to the right level.

Q: What's the advantage of using two boilers?

A: Each boiler is sized to half of the maximum load. For instance, let's say the total required load on the coldest day of the year is 250,000 BTU/Hr. If we use two 125,000 BTU/Hr. boilers instead of one 250,000 BTU/Hr. boiler, we'll be burning about half as much fuel during a large part of the heating season.

Q: You said we'll be getting rid of the open expansion tank in the attic when we convert the system to forced circulation. Why do we have to do that?

A: Cast-iron and steel boilers last a lot longer when the system is closed. That's because there's a lot less oxygen corrosion in a closed system.

Q: Do I *always* have to get rid of the open tank?

A: Not necessarily. A good choice for a gravity conversion job is a copper finned-tube boiler. These boilers are non-ferrous and deal with oxygen especially well. They're also impervious to thermal shock (they have flexible heat exchangers) and work well with cooler water (typically down to 105°F).

Q: Let's say I decide to close the system. What do I need to know to size a closed compression tank for a conversion job?

A: You'll have to know three things:

1. The gallons of water in the system
2. The difference between the fill and relief pressures, and
3. The average water temperature of the system, which in this case should not be more than 170°F.

Q: Why is the average water temperature limited to 170°F?

A: So the water won't flash to steam in the open attic tank. The old-timers sized their radiation to provide plenty of heat on the coldest day of the year with a maximum high limit of 180°F at the boiler. The water would leave the boiler at 180° and return at about 160°, giving them an average temperature of 170°F within the radiation.

Q: If I run the system with hotter water what will happen?

A: You'll probably overheat the people and raise their fuel bills.

Q: What are the guidelines for sizing a plain steel compression tank for a gravity conversion job?

A: Measure the total system radiation and then apply this rule of thumb:

1. *When there is less than 1,000 square feet of radiation on the job, multiply the total by .03 to determine the tank size in gallons.*
2. *If the total radiation is between 1,000 and 2,000 square feet, use .025 as a multiplier.*
3. *If the total radiation load is greater than 2,000 square feet, use .02 as a multiplier.*

This will give you the size of the standard steel compression tank in gallons.

Q: How will I know how many square feet of radiation each radiator contains?

A: You can use this chart as a guide:

Thin-Tube (Water-Type) Radiators

	Height (Inches)	Sq. Ft. Per Section		Height (Inches)	Sq. Ft. Per Section
Three-	20"	1.75	Five-	20"	2.66
Tube	23"	2	Tube	23"	3
	26"	2.11		26"	3.5
	30"	3		32"	4.33
	36"	3.5		37"	5
Four-	20"	2.25	Six-	20"	3
Tube	23"	2.5	Tube	23"	3.5
	26"	2.75		26"	4
	32"	3.5		32"	5
	37"	4.125		37"	6
			Seven-	13"	2.625
			Tube	16½"	3.5
				20"	4.25

Q: What is a Square Foot of Equivalent Direct Radiation equal to in BTU/Hr.?

A: For gravity conversions, we can say that each square foot of EDR will be equal to 150 BTU/Hr. when the average water temperature is 170°F.

Q: Will these tanks be larger than they would be on a more-modern system?

A: Yes, these tanks will be a *lot* bigger than the ones you'd use on a job sized for forced-circulation. That's because jobs sized with circulators in mind use smaller pipe. Smaller pipe means less water in the system. Less water means less expansion, and less expansion means a smaller compression tank.

Q: Suppose I want to use a diaphragm-type compression tank, how should I size it for a gravity conversion job?

A: You can use this rule of thumb:

*Take the size of the standard steel compression tank in gallons and multiply by .55 if the building is two stories tall or .44 if the building is three stories tall. The answer will give you the **volume** of the diaphragm tank.*

Q: Can you give me an example of this?

A: Sure! Let's say we have a two-story house with 1,000 square feet of radiation. We'll size a standard steel tank first: 1,000 x .03 = 30 gallons. Now, since it's a two-story house, we have to multiply that by .55 to get the *volume* of the diaphragm tank. (30 x .55 = 16.5 gallons of required volume in the diaphragm tank)

Q: Where do I find the "volume" of the diaphragm tank?

A: In the manufacturer's specification sheets. Here, for instance, are the volume ratings of standard diaphragm-type tanks made by Amtrol, Inc.:

Model Number (Amtrol)	Volume (in gallons)
15	2.0
30	4.4
60	7.6
90	14.0
SX-30V	14.0
SX-40V	20.0
SX-60V	32.0
SX-90V	44.0
SX-110V	62.0
SX-160V	86.0

And here are the volumes of the tanks made by Vent-Rite (Flexcon Industries):

Model Number (Vent-Rite)	Volume (in gallons)
VR 15 F	2.1
VR 30 F	4.5
VR 60 F	6.1
VR 90 F	21.0
SX VR30 F	21.0
SX VR40 F	21.0
SX VR60 F	29.0
SX VR90 F	37.0
SX VR110 F	53.0
SX VR160 F	74.0

For the building in our example, you'd use either an Amtrol

SX-40V, a Vent-Rite VR 90 F, or any combination of smaller tanks which equal or exceed 16.5 gallons of volume. If you wanted, you could use four Amtrol 30s or four Vent-Rite VR 30 Fs, for example.

Q: Do I need to check anything on these tanks before I install them?

A: Yes, always check the air pressure on the diaphragm side of the tank. It should equal the system fill pressure **when you have the tank disconnected from the system.** The fill pressure for a two-story building is typically 12 psig; for a three-story building, it's 18 psig. If the pressure is too low, use a bicycle pump or an air compressor to increase it. The tank's pressure (when disconnected from the system) should always equal the system fill pressure (the pressure reducing valve's setting).

Q: What method should I use to size the replacement boiler?

A: You should size the replacement boiler based on two things: an accurate heat loss calculation of the building, and an accurate measurement of the existing radiation. Don't settle for one or the other, check them both and compare.

Q: Why is this so important?

A: By checking both the heat loss and the radiation, you'll be able to calculate the proper design temperature for your converted system. Many old-timers oversized their radiators because the radiation charts available at the time listed only steam ratings. One square foot EDR in steam work will put out 240 BTU/Hr. One square foot EDR in hot water work (based on an average water temperature of 170°F) will put out 150 BTU/Hr. This is because water at 170°F is cooler than steam at 215°F.

To compensate for the charts, the old-timers added 60 percent to their radiation sizing. This, as you can imagine, led to some hefty oversizing.

Q: Is that a bad thing?

A: It can actually work out to be a *good* thing. If the radiators are oversized, you'll be able to run the system at a relatively low average water temperature. I've found that most conversion jobs run well at an average water temperature of 150°F (in the

New York City area), and that's on a day when the outside temperature reads zero! Lower boiler water temperatures mean
lower fuel bills.

**Q: Is there ever a time when I should oversize the new boiler
on these jobs?**

A: No! There is absolutely no reason to oversize a boiler. Base
the size on the heat loss of the building **as it exists today**. Pipe
it properly, using the bypass line we discussed before. Then, if
the job is overradiated, lower the high limit water temperature
accordingly to save fuel.

Q: What sort of hydronic accessories do I need on these jobs?

A: Use a good air separator to limit the possibility of air noises
and lack-of-heat problems. Locate it in your new near-boiler
piping on the supply side of the system (where the water is
hottest), just before the circulator. You should locate the compression tank at a point near the air separator.

Fill the system with a pressure-reducing valve at the point
where you've connected the compression tank into the system.
That's the "point of no pressure change," the only place in the
system where the circulator's pressure can't affect the system's
pressure.

You'll also need a flow-control valve to prevent gravity circulation when the circulator is off. Pipe it in right after the circulator.

Q: If I wanted, could I switch the system back to gravity?

A: Yes, that's one of the nice things about these conversion
jobs. They're real easy to switch back (temporarily, at least)
should something happen to the circulator. All you have to do
is crank open the little lever at the top of the flow-control valve
and the hot water will once again rise out of the boiler and up
into the radiators.

Q: What are my control options with these conversion jobs?

A: Well, there's primary/secondary pumping. We looked at that
before. You can also install thermostatic radiator valves on the
radiators.

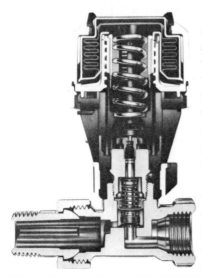

These devices sense the air temperature in each room and modulate the flow of water through the radiator. They're completely self-contained and need no electrical wiring. They last for years, are relatively inexpensive and have been around since the 1920s. I've found they keep the room temperature within one or two degrees F of the set point. With thermostatic radiator valves, every room becomes its own zone.

Should you choose to use them, set the circulator to run continuously during the cold months. The valves will take care of the comfort level in each room. If you want to take control a step further, vary the boiler temperature based on that outdoor-air reset controller I mentioned before. This control also helps do away with any expansion/contraction noises you may have in the system.

Q: Is there a simpler way to control a conversion job?

A: The simplest way is to have the house thermostat turn on the burner and the circulator at the same time. This doesn't give you the ability to zone each room, but it is less expensive, and it works. Don't forget that bypass line around your new boiler.

Q: Suppose I decide to keep the old boiler and just add a circulator and a flow-control valve. Will this save me fuel?

A: Don't be surprised if it *increases* the fuel bills! Old boilers and gravity systems work well together because when the burner shuts off, the residual heat in the boiler rises up into the radiators. However, when you install a flow-control valve, the residual heat goes up the flue instead of into the radiators. The result? Higher fuel bills.

Q: How about if I just install a circulator on that old boiler and forget about the flow-control valve?

A: This will help lower the fuel bills by moving the hot water to the radiators more quickly, while not stopping the residual heat from moving into the radiators. You'll have to fiddle with the thermostat's heat anticipator, though, to keep the system from overriding. Also, you may need more than one circulator if there's more than one set of supply and return lines.

Q: **Can I add a zone to an existing gravity system by tapping into the supply and return lines with a circulator and a loop of baseboard?**

A: I wouldn't do this. The forced flow through your new zone is bound to affect the operation of your gravity system. How it affects it will vary from system to system (no two are alike), but from what I've seen, it usually leads to problems. I'd avoid it if I were you.

If you're interested in zoning, think about adding a circulator to the main part of the house as well, and consider those thermostatic radiator valves I told you about before.

Q: **Were there specialized systems of gravity hot water heating?**

A: Yes, Honeywell made a system called "accelerated hot water heating" which was very popular in its day.

Q: **When did they use this system?**

A: In the early days of this century.

Q: **Are these systems still around?**

A: There are enough of them out there to make you scratch your head in wonder.

Q: **What was Honeywell trying to achieve with this system?**

A: They wanted to find a faster way to move the water from the boiler to the radiators. They knew that if they could do this, they would save the consumer money on fuel.

Q: **Why didn't they just use a circulator?**

A: Because circulators hadn't been invented yet!

Q: **So how can you make the water move more quickly with-**

out using a circulator?

A: By raising its temperature. The hotter the water, the faster it flowed.

Q: But if they raised the water temperature, wouldn't there be a problem with water boiling in the open expansion tank?

A: Yes, under normal circumstances, but with the Honeywell system, the old-timers were able to run the system under pressure.

Q: How much pressure?

A: As much as 10 psig at the top floor, and since the boiling point of water increases with a rise in pressure, they could have temperatures as hot as 240°F in the radiators. That made the water circulate very quickly.

Q: Was there a danger in pressuring this type of system?

A: Normally, there would be because the expansion tank was the weak link. It was usually made of copper or galvanized steel and held together with rivets. It wasn't built to take the strain. Apply too much pressure and the tank could (and frequently did!) explode, taking the top of the house with it.

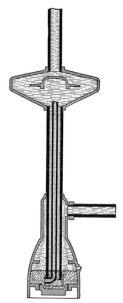

However, with the Honeywell system, a special device called a Heat Generator kept the tank separated from the boiler, the system piping and the radiation.

Q: What did this device look like?

A: It was made of cast-iron and stood about 2½ feet high.

There was a narrow steel tube inside the unit's main tube, and it dipped down into a pot filled with mercury.

Q: Why did they use mercury?

A: Because it's heavy. They used the mercury to separate the water in the boiler, piping and radiation from the water in the open expansion tank. Here, take a look at how the Heat Generator tied into the system.

The top pipe went up to the open tank. The side pipe connected the system to the Heat Generator. The mercury kept the two sides separated.

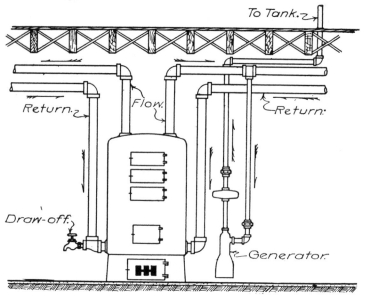

Q: How did the Heat Generator work?

A: As the old-timer built pressure on the system, the water in the boiler, piping and radiation expanded and pressed down on the mercury.

The mercury rose up the narrow tube and cascaded back down into the pot through the wider, outer tube. As long as the water was expanding, the mercury continued to circulate.

Q: Why didn't the mercury rise up into the open expansion tank?

A: Because of its weight. Mercury's pretty heavy. In fact, it's nearly fourteen times heavier than water.

Q: Can the water in the boiler, piping and

radiation enter the bottom of the mercury tube?

A: Yes, if the system pressure rises high enough. The water will then enter the tube and separate from the mercury in that wide separation chamber at the top of the Heat Generator. From there, it rises into the expansion tank.

Q: So the Heat Generator wouldn't let the system pressure rise above a certain point?

A: Right! It limited the system pressure to 10 psig at the top, without putting any pressure at all on the open expansion tank. This made the operation thoroughly safe, and it also made the water circulate very rapidly.

Q: I can see how Honeywell's device increased the speed at which the system heated, but what advantage, if any, did it offer the installer?

A: Because of the higher temperatures, the installer could downsize all his radiation by as much as 15 percent.

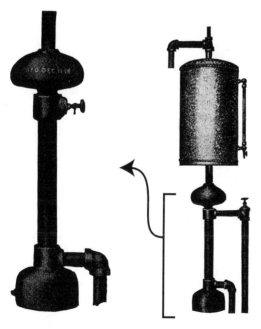

Q: Did the old-timers use other types of devices such as this one?

A: Yes, there was a similar one called the Klymax Heat Economizer (sounds sexy, doesn't it?). Here's a picture of one attached to the bottom of an open expansion tank.

Q: Were there any others?

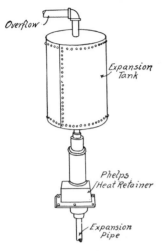

Overflow

Expansion Tank

Phelps Heat Retainer

Expansion Pipe

A: There was also the Phelps Heat Retainer.

This device operated by opening and closing a double-acting valve that was encased inside a cast-iron box. The side of the valve that opened to the atmospheric tank held a 16½ pound weight. This weight would lift and open the valve when the system reached 250°F. The expanded water then moved safely into the open tank.

When the pressure dropped below 16½ psi, the weight closed the valve and the shrinking water opened a retainer valve that allowed the water in the tank to reenter the system piping.

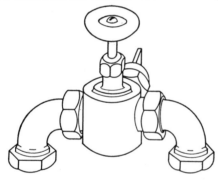

Q: **Did Honeywell use a special valve at the radiators?**
A: Yes, they had something called the "Unique" valve, and from the looks of it, I'm sure you can see why they called it unique!

Q: **How did this valve work?**

A: To understand, you have to look inside. Here's a picture of the valve when it's closed.

As you can see, water flows past the radiator when the valve is in this position, but look what

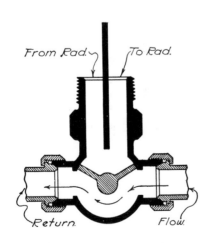

From Rad.   To Rad.

Return   Flow

happens when you open the valve.

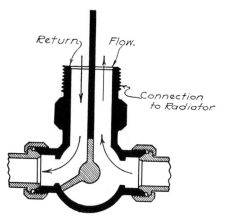

The water now enters the radiator on one side of the internal baffle, as the cooler, return water moves in counterflow past the other side of the baffle.

Q: Was this the same Honeywell Company we know today?

A: One and the same!

CHAPTER TWO:

❧

INDIRECT
HOT-WATER
HEATING

❧

*"Dan, you gotta see this house!
They got boilers hanging
inside the ductwork!"*

—A bewildered heating
contractor, circa 1989

Q: What's "indirect" heating?

A: It's a method of heating the old-timers used to warm fresh air before it entered a building.

Q: How did they do this?

A: They placed the indirect radiators outside the room they were heating (that's where the name "indirect" comes from). They used ductwork to direct the fresh outside air across these remote radiators where it would be warmed before rising into the rooms.

Q: Where exactly did they place these radiators?

A: Almost always in the basement, and usually right under the room they were heating, but they could also put them in a central location and route the ductwork off to the rooms. It all depended on the size and shape of the house.

Q: What sort of houses had this type of heating system?

A: Big houses! Indirect heating was mostly the province of rich folks. Rich folks loved this type of heating system.

Q: How come?

A: At the turn of the century many people believed stale air (they called it "vitiated air") was a prime cause of disease. Those who could afford it heated their homes with a constant supply of warmed fresh air. They thought this was the healthiest heating system possible.

Q: Were they right?

A: Since this was the first residential, central heating system to combine heating with ventilation, they were certainly on the right track. Indirect hot-water heating was an improvement over the direct hot-water and steam systems of the time, and as far as health goes, it was a great improvement over the stoves most people used.

Q: Were these systems expensive to operate?

A: There sure were! That's why you'll find them mostly in the homes of long-gone wealthy folks.

Q: What does an indirect radiator look like?

A: Most look like a small cast-iron sectional boiler, without the jacket, of course.

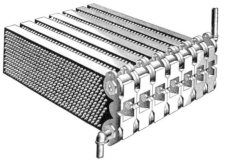

The old-timers used iron rods to suspend them from the basement ceiling inside the duct. Many are heavy enough to cause the floor above to sag. The first time you run across one can be a pretty intimidating experience.

Q: Were all the radiators made from cast iron?

A: No, some were steel, and these looked a lot like fin-tube radiation, minus the enclosure.

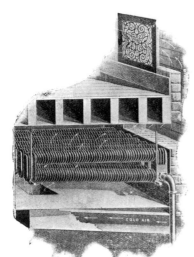

A typical steel indirect radiator could have a dozen or more tubes set up side by side within the duct. Air flowed by gravity from the outside to the rooms.

Q: Did they heat the entire house with these?

A: Sometimes they did, but most often they used a combination of indirect radiators and direct radiators. Direct radiators are the "regular" ones that sit in the room and heat the same air over and over again.

Q: How did they decide where to put the direct radiators and the indirect radiators?

A: The large rooms on the first floor were almost always heated with indirect radiators, as were the bedrooms on the second floor (they worried about vitiated air getting them as they slept). If the house had a third floor, it was usually heated with direct radiation since these rooms typically were not used every day. The servants' quarters were always heated with direct radiation.

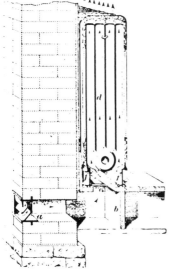

Q: Were there any radiators that heated both directly *and* indirectly?

A: Yes. These were "regular" radiators that the old-timers set over an outside air duct. Here's what one looks like.

They did a nice job of mixing and warming fresh air from the outdoors with the air that was already in the room.

Q: Did these direct/indirect radiators heat by convection?

A: Yes, they did. Whenever you place a radiator directly in the room you'll get convection currents. The heated air will rise off the radiator as the cooler air scoots across the floor.

Q: Did indirect systems promote convective air currents in the room?

A: Not to the same extent because the heat source was the floor or wall register. Return air had no way of flowing into a cooler place.

Q: So where did the return air go?

A: It left the house through cracks around the windows and doors.

Q: How can you tell the potential BTU/Hr. output of a cast-iron indirect radiator?

A: First, you have to be able to see the radiator well enough to count and measure its sections. The duct will have an access door, but it's sometimes difficult to get a clear view of the radiator.

Q: Let's say I can get the measurements I need. Where do I

find the ratings?

A: The American Radiator Company was once the largest manufacturer of indirect radiation. The unit you have is probably one of their "Vento," "Perfection Pin," or "Sanitary School Pin" models. Their 1925 book, *The Ideal Fitter*, shows these ratings:

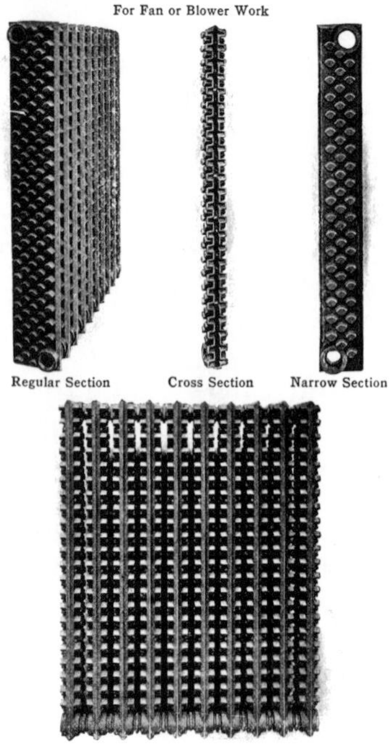
For Fan or Blower Work
Regular Section
Cross Section
Narrow Section

Vento Indirect Heaters:

Narrow Sections	Sq. Ft. per Section	Height	Width
40 inch	7.5	41 1/64"	6 3/4"
50 inch	9.5	50 29/32"	6 3/4"
60 inch	11.0	60 11/16"	6 3/4"

Regular Sections	Sq. Ft. per Section	Height	Width
30 inch	8.0	30"	9 1/8"
40 inch	10.75	41 1/64"	9 1/8"
50 inch	13.5	50 29/32"	8 1/8"
60 inch	16.0	60 11/16"	9 1/8"
72 inch	19.0	72"	9 1/8"

Perfection Pin

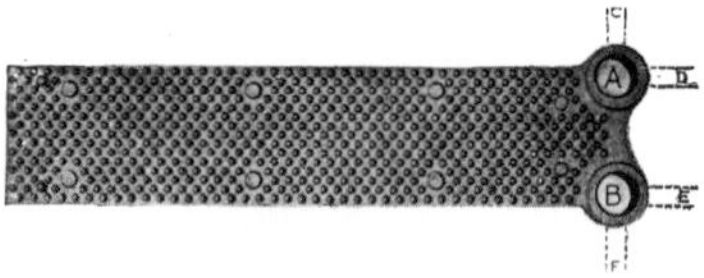

Measurements: Length, 36 1/4 inches. Height 7 1/2 inches—at connecting point, 9 13/16 inches. Width in stack, 2 3/4 inches.

Heating surface (per section): 10 square feet.

Sanitary School Pin

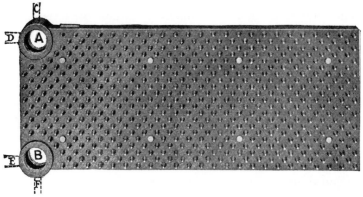

Measurements: Length, with regular tappings, 36⅛ inches—when tapping is at "D" or "E," 36⅜ inches. Height, 13⅞ inches; height at connecting point with regular tappings, 15¼ inches—with special tappings at "C" or "F," 15½ inches. Width—each section in stack, 4 inches.

Heating surface (per section) 20 square feet.

Q: What is a Square Foot of Equivalent Direct Radiation equal to in BTU/Hr.?

A: For indirect radiators, we can safely say that each square foot of EDR will be equal to 150 BTU/Hr. when the average water temperature is 170°F.

Q: Okay, now suppose I can't see the radiator at all?

A: As I said, there's almost always an access door in the duct-work, but if worse comes to worst, you can estimate the size of the indirect radiator by checking the size of the hot water tapping feeding it. Here's a rule of thumb for you:

*A 1¼" tapping can feed up to
60 square feet of indirect radiation.*

*A 1½" tapping can feed up to
90 square feet of indirect radiation.*

*A 2" tapping can feed up to
100 square feet of indirect radiation.*

Q: Can I check the size of the radiator by measuring the size

of the hot air flue connected to it?

A: You can get a fairly good idea. The old-timers based the size of the hot-air flue on the square foot of connected indirect radiation that it contained. For hot-water jobs, they allowed two square inches of flue per square foot of radiation.

If you have absolutely nothing else to go by, figure the square inches of the hot-air flue (by multiplying its two sides together) and then divide that by 2. That will give you a good estimate of the square feet of radiation inside that duct.

Q: **How about the cold-air inlet flue, what size will that be?**

A: Usually, it's between ⅔ and ¾ the size of the hot-air flue.

Q: **Why is the cold-air flue smaller than the hot-air flue?**

A: Because hot air takes up more space than cold air.

Q: **Does it matter where the old-timer set the radiator inside the duct?**

A: Yes, an indirect radiator should be about 10 inches below the top, and at least eight inches above the bottom of the flue. It should also be tight against the sidewalls of the duct.

Just like this.

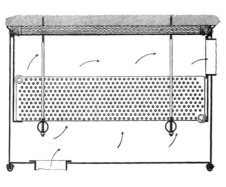

Q: **Are these dimensions crucial?**

A: They are if you want to get the proper flow of air across the radiator.

Q: **But isn't this something the old-timers took care of years ago? Why should I have to worry about it now?**

A: The old-timer *did* take care of it many years ago. But suppose that cast-iron indirect radiator you're staring at decides to leak tomorrow? You can't buy a new one, so you might have to replace it with a homemade nest of steel fin-tube radiation. *That's* when you have to pay close attention to those dimensions.

Q: Does it matter which way the air approaches the radiator?

A: The old-timers always tried to get the cold air to come in from the bottom of the radiator.

Set up this way, they could take advantage of natural convection currents. If they couldn't do it this way, they took their next best option and brought the fresh air in from the side opposite the warm air outlet.

Q: Are the registers in the rooms the same size as the flues that feed them?

A: No, the registers are usually about 25% greater in area than the flues.

Q: Did they use fans to move the air around the house?

A: No, this system worked purely on convection currents. Hot air, being lighter than cold air, naturally rises up into the building.

Q: Does the hot air move more quickly to the top floors than it does to the lower floors?

A: It sure does, and that's because of the chimney effect created by the taller, second- and third-floor flues.

Q: How much faster does the heated air move to the upper floors?

A: Typical air velocities are 1½ feet per second to the first floor, 2½ feet per second to the second floor, and 5 feet per second to the third. As you can see, the air speeds up quite a bit as it moves higher.

Q: So could the old-timer feed more than one floor with a single flue?

A: No, because of these differences in velocity, each flue served only one floor.

Q: Did they size the flues going to the upper floors differently from the way they sized the flues leading to the lower floors?

A: Well, they had to size for BTU load, of course, but given that two rooms (one on the first floor, the other on the second or third) were equal in size, the flue going to the upper floor could be about 25% smaller than the one serving the lower floor because of the greater air velocity.

Q: Were the registers on the upper floors smaller than the ones on the lower floors as well?

A: Yes, they were. And this is where things can get tricky if all you're looking at is the register.

Q: Can I equate modern register or duct sizing with this type of heating?

A: Not if you want to stay out of trouble. The two are **very** different.

Q: Did they use a circulator to move the water from the boiler to the indirect radiators?

A: No, they didn't have circulators in those days. They used the same piping techniques we looked at before when we talked about gravity hot-water heating. The difference here is that the runs to and from the radiators aren't as long as they are in a direct, gravity hot-water system. That made the piping less complicated to size and install.

Q: How did the old-timers figure the heat loss for these old buildings?

A: Most used the "Mills Rule."

Q: What's that?

A: It was a rule-of-thumb method developed by John Mills, a famous 19th Century heating engineer. Simply put, the Mills Rule allowed:

One square foot of radiation for each
2 square feet of glass in the building

One square foot of radiation for each
20 square feet of exposed
(not heated on the other side),
wall, ceiling or floor in the building, and

One square foot of radiation for each
200 cubic feet of room volume.

The old-timer would total these three things, and that would be the heat loss for the building.

Q: Can you give me an example of the Mills Rule in action?

A: Sure! Let's take a room measuring 10' x 20'. Two walls are exposed to the outside and the ceiling is 10'. The room is on the first floor, under a heated second floor, but over an unheated basement. In one of the walls there's a window that measures 5' x 4' and in the other wall there's a window measuring 8' x 5'.

First, let's figure the glass. The 5' x 4' window equals 20 square feet, and the 8' x 5' window gives us another 40 square feet for a total of 60 square feet of glass. Divide that by 2 to convert it to square feet of radiation and you'll come up with 30 sq. ft. of radiation required to overcome the heat loss through the glass. If the room had a door, you'd treat it the same way.

Now for the "walls." There are three exposed surfaces in this room—the two walls and the floor. We'll treat them all as "walls." We don't consider the ceiling because it's under a heated space so there will be no heat loss there. The same goes for the other two walls; they're also next to heated spaces.

Let's convert the two walls and the floor to square feet:

10' x 20' = 200 sq. ft. (first wall)

10' x 10' = 100 sq. ft. (second wall)

10' x 20' = 200 sq. ft. (the floor)

We have a total of 500 square feet of surface exposed to the cold outdoors. Divide that by 20 to get square feet of radiation and you'll come up with 25.

Next, we measure the volume of the room by multiplying the length by the width by the height: 20' x 10' x 10' = 2,000 cubic feet. Divide that by 200 to get square feet of radiation. The answer is 10.

Now you add the glass (30 square feet of radiation) to the walls (25 square feet of radiation) to the volume (10 square feet of radiation). The total is 65 square feet of radiation.

Q: Was the Mills Rule based on steam or hot water?

A: It was based on steam. Radiation was only rated for steam when John Mills was alive. To figure their hot-water radiators, the old-timers would add 60% to the Mills numbers.

Q: So for the example problem, how many square feet of hot water radiation would they have used?

A: They would have taken the 65 square feet they came up with based on the Mills Rule and multiplied it by 1.6. That would give them 104 square feet of hot water radiation.

Q: What is that equal to in BTU/Hr.?

A: Each square foot of hot water radiation is equal to 150 BTU/Hr. This is based on an average water temperature of 170°. So according to the Mills Rule, the heat requirement for the example would be: 104 x 150 = 15,600 BTU/Hr.

Q: Will this method of calculating heat loss work today?

A: Not with our modern construction methods. If you used the Mills Rule today, you'd wind up with a system that would be much larger than needed on the coldest day of the year. Your price would be way out of line if you were competing against someone who did an accurate heat-loss calculation.

Q: If I calculated the heat loss of this same room using modern computer software what would I come up with?

A: About 40% of what the Mills Rule gave us.

Q: How about if I'm figuring a heat loss on an old house? Would it be safe to use the Mills Rule then?

A: Only if the house has all the original equipment. In other words, no replacement windows or doors, no weather-stripping, no insulation. When I figured the heat loss on the sample room using the heat-loss software, I assumed someone had installed replacement windows and insulation. Makes a big difference, doesn't it?

Q: Considering the old-timers were using fresh air to heat the building, did they have to increase the load even further than what the Mills Rule called for?

A: They sure did, and by an impressive amount! Typically, for an indirect hot-water system they'd use the Mills Rule to figure the heat loss of the house as though it were going to be heated with direct steam radiators. Then, as I said, they'd add 60% to their radiation load to make it work with hot water, which is cooler than steam. *And then, they'd add a whopping 75% to what they already had to compensate for the cold fresh air that would be coming in through the basement ducts.*

Q: So in the case of the example room, what would happen?

A: The old-timers would have taken the 15,600 BTU/Hr. load and multiplied it by 1.75. Then they would have installed indirect radiation capable of dumping 27,300 BTU/Hr. into the room.

Q: Is that why those indirect radiators are so humongous?

A: Yep.

Q: Suppose I'm replacing a boiler and I find out the home owner is no longer using the 100% fresh air feature of the indirect system, do I still have to add that 75% extra to my load calculation?

A: No. If you did, your boiler would be ridiculously oversized.

Q: Is it practical to keep an indirect system operating nowadays with 100% fresh make-up air?

A: When you consider the cost of fuel, this system really isn't practical anymore. We've learned that "vitiated air," the scourge of Victorian America, is not the only thing that can make you ill. Many home owners, in an attempt to save money, have chosen to abandon the ventilation side of the indirect system.

Q: How do you abandon the ventilation side of the system?

A: Usually, all you have to do is seal the cold air inlet duct and open the radiator access door. That will allow the air in the basement to move through the heaters and up into the rooms.

Q: Does this always work?

A: I've seen mixed results. How well it works depends largely on whether or not the air in the rooms can work its way back down to the basement. Also, if the basement isn't squeaky clean, the smells that waft upstairs will have the home owners holding their noses. It's a guaranteed call-back for a contractor.

Q: Suppose the air in the rooms can't work its way back down to the basement?

A: You're going to get less heated air from the registers than you did when you were using the outside air duct, of course. Remember, air can't come out of the registers in the rooms unless air can get in at the heaters, so you may have to cut return air ducts into the first floor rooms to get good circulation.

Q: Is this practical?

A: Not always.

Q: Which brings up another thought. When the outside-air duct was open, how could all that fresh air enter the house? Didn't an equal amount of air have to leave?

A: Good question. As the warmed, fresh air came up through the registers, the air that was already in the rooms had to leave, and it did. It went out through the many cracks in the building, around the windows and doors, the places where floor and walls came together, the fireplaces. The house acted like a chimney.

Q: So if I were to weatherize and tighten up an old house that had this type of indirect heating system I could actually mess up the heating system?

A: That's right. If you stop the air from leaving, less air will enter and the heating capacity of the radiators will drop off considerably.

Q: But will I still be able to heat the house?

A: It depends. Most of the time, the old-timers were so generous with their radiation sizing that you'll probably find the house heats well even with the lesser movement of air through the indirect radiators.

Q: So I won't have a problem?

A: Hey, I never said that! I remember visiting a house with indirect hot-water heat once. The home owner wanted to save fuel so he sealed all the cracks in the house and closed and caulked the outside air duct. Not too much air got in or out of his house. And because it couldn't, he was freezing. And yet his fuel bills were higher than ever because no warm air could make it out of the basement ductwork and up to his thermostat. The burner ran all day, and yet the guy was freezing. Imagine that.

CHAPTER THREE:

DIVERTER-TEE HOT-WATER HEATING

"When I bleed the radiators, I don't get any air. How come?"

"Cause it ain't an air problem!"

—A living-room-floor conversation, circa 1980

Q: What's the principle of the diverter-tee system?

A: This system lets you supply hot water to a radiator and return cooler water from the same radiator by using a single pipe.

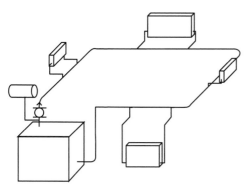

Q: Is this why everyone calls it a "one-pipe" system?

A: Yes, the supply and return main are one and the same. Other types of hot water systems use one pipe to supply hot water to a radiator and a second pipe to return the cooler water from that same radiator to the boiler. We call these systems "two-pipe."

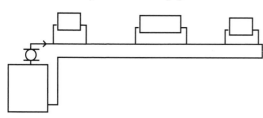

Q: How did the diverter-tee system come to be?

A: Back in the days when most folks used steam to heat larger buildings, the people who sold hot water equipment had a problem. How can you sell a heating system that requires two pipes when your competition (steam) only needs one? The answer was the one-pipe diverter-tee system.

Q: Was this system widely accepted by contractors?

A: It sure was! Compared to steam, the one-pipe diverter-tee system cost less to install, was quieter in operation and used much smaller pipes.

Q: Which manufacturers promoted this type of system?

A: Primarily Bell & Gossett. They brand-named their fittings "Monoflo" and published a series of popular handbooks during the Thirties and Forties that explained how to size and install these systems. Taco also did a good job of promoting what they called a "venturi" fitting. For our purposes, however, we'll overlook the manufacturers' brand-names and just lump them together as "one-pipe diverter tees."

Q: Were Bell & Gossett and Taco the first to come up with the idea of a diverter tee?

A: No, the credit for that goes to Oliver Slemmer, a Cincinnati, Ohio heating engineer who designed and patented his O-S fitting in the early days of this century. Here's what the O-S fitting looked like.

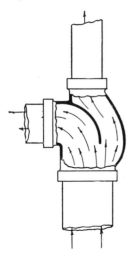

Q: Where would I find O-S fittings?

A: You'll find them on old gravity hot-water heating systems, the kind we went over in Chapter One.

Q: Where will I find the more-modern, one-pipe diverter-tee systems?

A: Mainly in buildings constructed during the 1940s and 1950s. This system was very popular with home builders of the time because it offered them the benefits of hydronic heat without the labor-intensive aspects of steam- and gravity-hot-water heating.

Q: How does the diverter tee work?

A: To answer your question, we first have to take a look at what goes on when water enters a *standard* tee.

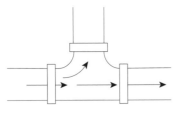

The water goes in one way, but it can come *out* in an infinite variety of ways. For instance, if four gallons per minute entered this tee from the left side, *any* combination of flows totaling 4 gpm would be possible through the two outlets. The flow could split in half with 2 gpm flowing through the side of the tee and 2 gpm flowing down the run. Or 3 gpm could go straight while 1 gpm traveled out the side. Anything is possible with a tee.

Q: Could all 4 gpm go straight through the tee with no flow whatsoever going through the side?

A: Sure! All it would take to make that happen would be a partially closed valve (or anything else that creates a restriction) in the branch piping.

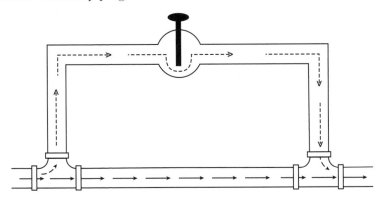

The valve increases the resistance to flow through the side of the tee, making it difficult for water to flow that way. So most of the water goes straight instead.

Q: Is this what we mean when we say water follows "the path of least resistance?"

A: Yes. Water will only flow from a point of high pressure to a point of low pressure. The greater the difference in pressure between those two points, the greater the flow will be. Anything that resists flow will lessen the difference in pressure between the two points, and that causes less water to flow that way.

Q: What would happen if that partially closed valve were in the piping *between* the two tees instead of in the branch piping?

A: You mean if you set the piping up like this?

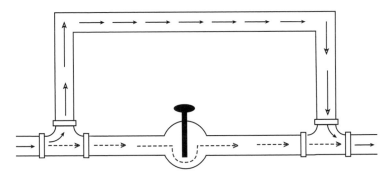

Well, then the resistance to flow along the main would increase. More water would flow through the branch piping and less water would flow along the main. The water will *always* follow the path of least resistance.

Q: Suppose you had your piping set up that way with two standard tees and *no* valves. Would any water flow through the branch piping then?

A: You mean like this?

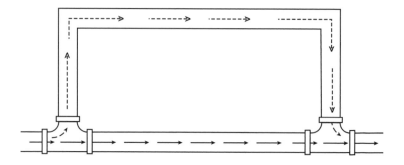

Piped this way, some water might flow through the branch, but my guess is most of the water would follow the path of least resistance and flow along the main. How come? It's just easier for it to go that way.

Keep in mind water won't necessarily flow through a pipe just because you have it connected to another pipe.

Q: How much water, if any, *would* flow through the branch in this case?

A: It depends entirely on the difference in pressure between the two standard tees.

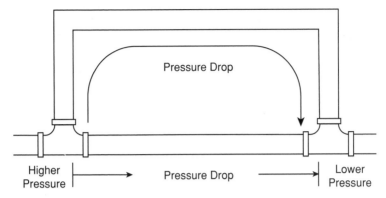

The greater the difference in pressure between these two points, the more water will flow through the branch. If there's hardly any difference in pressure between the tees, very little water, if any, will flow through the branch.

Q: Is there usually much difference in pressure between two standard tees?

A: Not if they're placed closely together. The only difference would be the friction the water creates as it flows along the main between the two tees. The closer you place the tees, the less friction, and the less pressure difference, there will be. That's why we need diverter tees to make this type of one-pipe system work.

Q: Wait a minute. Couldn't I just reduce the size of the pipe between the two standard tees? That would increase the friction in the main and create more of a difference in pressure between the two tees. Would that work?

A: Sure it would.

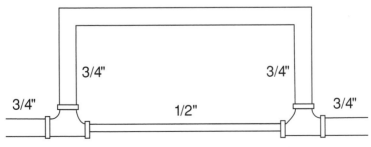

You know why? Because some of the water will find it easier to go through the ¾" branch piping than it will through the ½" main.

Q: How much water will flow through the branch?

A: That still depends on the difference in pressure between the tees. It's very difficult to tell. A lot depends on the lengths of the pipe.

Q: Suppose, in this case, my branch circuit is very long. Will water still flow that way?

A: Maybe, maybe not. It *still* comes down to a question of how much difference in pressure you have between those two tees.

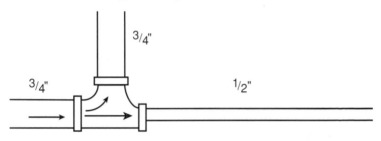

Let's look at the water's choices as it enters the tee on the left. It can go straight and put up with the pressure drop of the ½" pipe between the two tees. Or it can enter the branch and deal with the pressure drop the long branch piping offers.

Q: But wait a minute, the branch piping is *wider* than the main piping. Why should there be more pressure drop going that way?

A: It's because of the relative lengths of the two paths. There's more to this than just the width of the pipes, you know. The branch pipe is wider, sure, but it's also *longer*. The main pipe, on the other hand, is smaller by a full size than the branch pipe, but it's also shorter.

It *still* comes down to a question of difference in pressure between the tees.

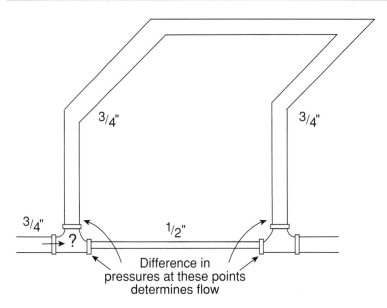

Q: Does this mean the length of the run to the radiator is important?

A: It's very important. I remember seeing a job where an installer connected nearly 100 feet of baseboard radiation to a single diverter tee off a one-pipe supply main.

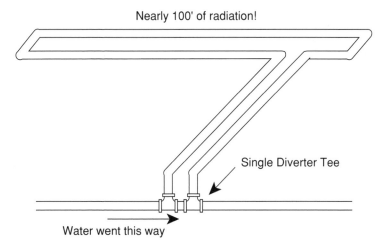

He couldn't understand why the baseboard wouldn't get hot. He was missing the key point—there was very little, if any, water flowing through the baseboard. And where there is no flow, there can be no heat.

Q: So just because I pipe a radiator into a main doesn't necessarily mean it's going to work?

A: Right, wishing and hoping will take you just so far. It *always* comes down to the differences in pressure between the tees and how much water flows through the main and through the branch. You have to think of flow as though it were a train on which heat travels like a passenger.

Q: So the diverter tee's job is to direct the flow into the branch?

A: Yes. It sets up a fixed resistance to flow along the main. The resistance drives a portion of the water into the branch. Here, take a look inside this one.

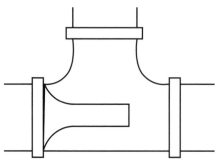

Diverter tees work because they change the water's path of least resistance.

Q: Why is that cone inside the tee?

A: The cone creates a narrow opening through which the water has to pass if it's going to continue flowing along the main. Since the cone makes it tough for all the water to go straight, some of the water will divert into the branch.

Q: Does it matter which way I face the cone?

A: Yes. If you're using one tee, it's best to put it on the return side with the wide end of the cone *receiving* the flow.

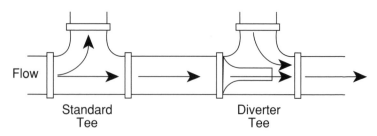

Flow

Standard
Tee

Diverter
Tee

Q: If the cone is on the *return* side, how can it divert the water up into the radiator?

A: You'll have to use your imagination to see what's going on here. Think of yourself as the water. You're flowing down the main toward the set of tees leading into the radiator. The first is a standard tee, the second, a diverter tee. You look up ahead and see a bottleneck on the main "road." That's the "traffic jam" the cone is causing. Things are slowing down so you decide to take the "service road" through the branch.

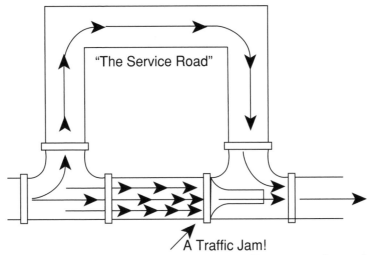

"The Service Road"

A Traffic Jam!

Once they're past the cone, the two flows—one from the main and the other from the branch—join again and continue to the next set of tees. Get it?

Q: **So the diverter tee's not "scooping" the water up into the radiator, is it?**

A: No, it's just creating resistance along the main. The water follows the path of least resistance through the branch.

Q: **Could I turn the diverter tee around so the cone points *toward* the flow and use it as the first tee instead of the second?**

A: You could, and it will work because the same principle applies: The diverter tee increases the pressure drop along the main and creates flow in the branch.

However, with the cone facing *into* the flow, the water will be a bit more turbulent and, if you're using enough tees, you might have to use a larger circulator to overcome the additional pressure drop caused by the turbulence. That's why most old-timers put the diverter tee on the return side with the wide end of the cone facing the flow.

Q: Do I need one or two diverter tees?

A: It depends on how much pressure drop you have to apply along the main to get the flow you need through the branch. As a rule of thumb, you should use one diverter tee if the radiator is above the main (using it as the return tee) and two diverter tees if the radiator is below the main.

Q: Why do I need two tees if the radiator is below the main?

A: Because hot water is buoyant and doesn't want to go down. The second diverter tee increases the pressure drop along the main and helps coax the hot, buoyant water down into the radiator.

Q: Suppose I use only one diverter tee to feed a radiator that's below the main. What might happen?

A: Most likely, you won't get the flow you need through that radiator. And you probably won't get the right amount of heat out of the radiator on the colder days of the year.

Q: I'll only notice the problem on the colder days of the year?

A: Probably, because we all size radiators to overcome the *worst-case* heat-loss condition. On milder days, the radiator is, in effect, oversized. The heat you'll get from it is enough to make the room comfortable on a mild day, but on a really cold day, the flow won't be there to bring enough heat to the radiator.

Q: Will venting the radiator help?

A: No, because this is a flow problem. It ain't an air problem.

Q: But when I vent the radiator, it does get hot all the way across and seems okay for a while. What's going on?

A: By opening up the vent on the radiator, you're temporarily changing the water's path of least resistance. It suddenly finds it easier to flow into the branch and toward the radiator because the vent is wide open to the atmosphere. In fact, the flow comes down from *both* the supply and return pipes when you open the vent. See? It's still a matter of pressure differential, but with the vent wide open, the water moves from, say, 12-psig system pressure to atmospheric pressure. That's why you get a temporary increase in flow rate and a corresponding increase in heat from the radiator. But as soon as you close the vent, the situation goes back to "normal" and your radiator goes cold again.

Q: So with diverter-tee systems it's easy to confuse flow problems with air problems?

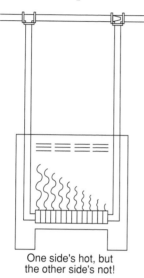

One side's hot, but the other side's not!

A: It's *very* easy to confuse the two. Any type of finned-tube radiation (baseboard or convectors) has the ability to give up a lot of heat in a short amount of space. If the flow rate is less than it should be, the front end of the radiator will be hot and the back end will be cooler. To the touch, it appears to be an air problem, but it's not.

Q: How will I be able to tell an air problem from a flow problem?

A: If you don't get air when you vent the radiator, it ain't an air problem! So stop venting.

Q: Is there a way of connecting free-standing radiators to minimize air problems?

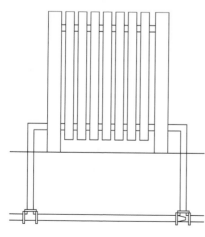

A: When the old-timers used free-standing radiators, they usually connected them bottom-and-bottom, like this.

Piped this way, water will continue to flow through the radiator, even if some air accumulates at the top.

Q: Will the water short-circuit through the radiator if I pipe it this way?

A: The hotter water will generally rise up into the radiator, displacing the colder water, because a free-standing, cast-iron radiator is a "wide space in the road." The flow through it is relatively slow so the hotter water rises.

Q: What happens if I pipe this radiator with one pipe at the top and the other at the bottom?

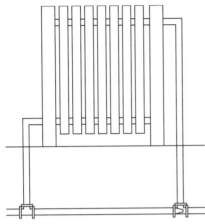

A: You mean like this?

Well, if you put either the supply or the return at the top, trapped air will be able to stop circulation through the radiator if enough gathers and creates an "empty space" up there at the top.

Q: How close together in the main can I pipe the two tees?

A: When the radiator is above the main, the tees can be as close together as six inches (if they're Bell & Gossett's tees) or 12 inches (if they're Taco's tees). If you have the radiator below the main, the tees have to be as wide apart as the ends of the radiator. This is very important.

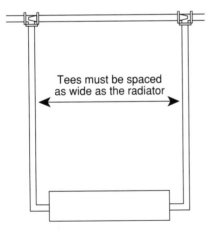

Tees must be spaced as wide as the radiator

Q: Why do the tees have to be so wide apart on a down-feed system?

A: Again, because the water in the main is hotter and more buoyant than the water in the radiator. It doesn't "want" to go down. By spacing the tees further apart, you increase the resistance to flow along the main and drive more water down through the radiator.

Q: Is the distance to my radiator important here?

A: Yes. Try to keep the piping to and from the radiator as direct as possible. By avoiding a lot of unnecessary fittings, you'll minimize the pressure drop going to the radiator and help establish circulation.

Q: Suppose I have a long run of baseboard radiation and it's below the main. How can I place those tees "the width of the radiation apart?"

A: It's difficult, isn't it? And this is a common problem many installers run into with these systems. Someone will replace an old convector with a long run of baseboard. He'll use the existing diverter tees, and then find the baseboard won't heat as well as the old convector did. There's too much resistance to flow through that long run of baseboard so most of the water bypasses the radiation and flows along the main.

Q: Does this one look like an air problem?

A: It looks like the worst air problem you've ever seen.

Q: But when I bleed the baseboard I don't get any air so it can't be an air problem, right?

A: Right, if you don't get air, it ain't an air problem.

Q: Can I solve the problem by using a larger circulator? Maybe a "high head" circulator?

A: If you try to jam more water down the pipe, you'll probably change the pressure differential relationship between the tees, and you may get a bit more flow through the radiator. But then you'll probably also get velocity noise and higher electricity bills. Here, as in most systems, finesse works better than brute force.

Q: So how can I solve this problem with the cold baseboard?

A: The best way would be to run the baseboard radiation as a separate zone if it *really* has to be that long to overcome the heat loss. Keep in mind, the heat loss of the room determines the amount of baseboard you need, not the length of the wall. Maybe you should check the length of the run before you do anything else. You know, see if you really need that much.

Q: Let's say I do need that much. Is there any other way around this problem of not having enough flow through the baseboard?

A: You could run the baseboard as a part of the main *within* the diverter-tee system.

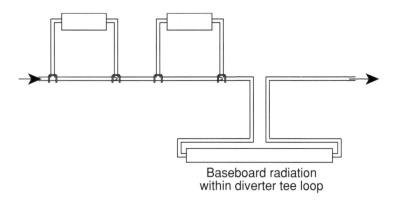

Baseboard radiation
within diverter tee loop

Make sure you cut out the two diverter tees and the main piping between the tees, and then just make the baseboard radiation a part of the main.

Q: Can't I just leave the tees in there?

A: No. With one of the outlet holes plugged, the tees will add too much resistance to the flow and may create insufficient flow problems everywhere.

Q: If I connect the baseboard as a part of the loop, can it be anywhere in the system?

A: It's best if it's the last thing on the loop.

Q: Why is that?

A: Because there's a good chance someone oversized the baseboard radiation, running it from wall-to-wall with no consideration of heat loss. If the baseboard is the last radiator on the line, the water will be relatively cool. It won't overheat the room as much as it would if it were the first thing on the line.

Q: If it were first on the line, could it cause any other problems?

A: If it's oversized and first on line, it might take too much heat out of the water. By the time the cooler water reaches the other radiators, it might not be carrying enough heat to keep the folks warm on the colder days of the year.

Q: Suppose I just want to get rid of a radiator. Do I have to get rid of the diverter tees, or can I just cap them off?

A: You can't close one of those outlets and expect to get the flow you were getting before. The pressure drop will be much greater.

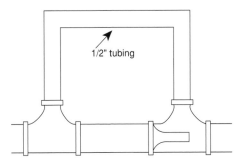

1/2" tubing

You can leave the tees in-line if you connect the branches with a ½" copper line. That gives the water somewhere to go and keeps things pretty much as they were before you cut the radiator off the line.

Q: **What's the best way to connect the radiators to the main if the system has radiators both above and below the main?**

A: Alternate the feeds and returns like this.

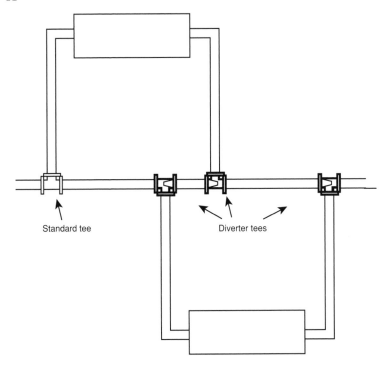

Standard tee   Diverter tees

Having the additional pressure drop of the upper radiator's diverter tee between the two down-feed diverter tees assists flow to the lower radiator. Vice versa for the upper.

Q: **Do I have to pitch the main in a diverter-tee system?**

A: If most of the radiation is below the main, you have to pitch the main at least one inch in twenty feet in the direction of flow. You'll also need an air vent at the end of the main to help you get rid of the air on start-up.

If most of the radiation is above the main, the pitch is not as crucial. The main, in this case, can be level, but a slight pitch in the direction of flow is always a good idea.

Q: Do I have to pitch the radiators as well?

A: It helps to pitch them up slightly in the direction of flow and to put an air vent on the return side of each one.

Q: When I'm troubleshooting, should I look for this pitch?

A: Yes, especially on the basement mains because people hang things on their basement heating pipes—laundry, lumber, children (They do pull-ups!), you name it. As you lose that one-inch-in-twenty-foot pitch, air problems develop.

Q: Is it difficult to get the initial air out of a diverter-tee system when you first start it up?

A: Yes, it usually is because the air goes to the top of the system where the water is coolest and the pressure is lowest.

Q: Is there any trick I can use to help me get rid of the start-up air?

A: Air will dissolve in water in proportion to its pressure. The higher the pressure, the more readily the air will go into solution. If you raise the system fill pressure to a point where it approaches the relief valve setting, more air will go into solution.

Q: What do you mean by "in solution?"

A: It doesn't appear as bubbles. It's invisible in the water and not a problem in the radiator.

Q: So, in other words, if I raise the fill pressure I'll have fewer air bubbles at the top of the system.

A: That's right, but you'll have to remember to reset the fill pressure to the proper setting after you've cleared the air. If you leave fill pressure at a too-high setting, your compression tank will be, in effect, undersized.

Q: When most of the radiators are below the main, I usually have a tough time getting rid of the start-up air. Are there any tricks that can help me with these?

A: There's one that works pretty well. When down-fed radiators won't heat because of air-related flow problems, most installers raise the water temperature. They believe the hotter water will move more quickly into the radiator. It seems to make sense— and that's why it doesn't work.

The trick, in this case, is to do what *doesn't* make sense. Instead of raising the water temperature, <u>lower</u> it. Then stand back and watch how much more quickly those down-fed radiators heat.

Q: How come that works?

A: Because the hotter the water is, the lighter it becomes. When you first start the system, the water in the radiators is cooler and heavier than the water that's circulating through the main. If trapped air is slowing the flow through the down-fed radiators, the water in the radiators won't have a chance to heat and become more buoyant. If you raise the water temperature, you aggravate the situation. But by lowering the water temperature, you'll bring the density of the water in the main closer to the density of the water in the radiators and have a much better chance of establishing circulation. Try it, you'll see what I mean.

Q: I've heard I should pipe my circulator so it pumps away from my compression tank. Is that true?

A: The circulator will *always* work best in this position, especially on these systems.

Q: How come?

A: Because air dissolves in water in proportion to the pressure applied to the system. When you pump away from the compression tank, you add the circulator's differential pressure to the system's fill pressure. When you pump toward the compression tank, you remove the circulator's differential pressure from the system's fill pressure. It's to your advantage to use the circulator's pressure every time the system runs. Trapped air bubbles will always go into solution more easily when the pressure is high.

Q: If I'm using a "high-head" circulator and I have it installed on the return, pumping toward the compression tank, will I have an even tougher time getting rid of the air?

A: You sure will. The higher the pump head, the greater the drop in pressure when it starts. Water-lubricated circulators can drop the system pressure in a typical house by about half the static fill pressure. The result is dramatic. You get air binding and air noise. Move the circulator to the supply side of the system, pumping away from the compression tank, and you'll see the difference immediately.

Q: How do diverter tees affect my circulator size?

A: Since this is a one-pipe system, all the water has to pass through all the tees. That means the pressure drop through diverter tees is cumulative. The more diverter tees you have, the more head pressure your circulator will have to develop.

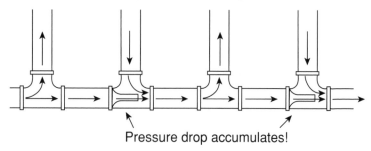

Pressure drop accumulates!

Q: How can I avoid having a circulator that's too large?

A: By splitting the loop.

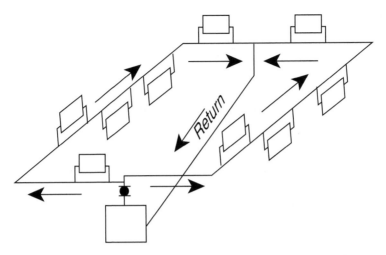

You have to size your circulator for the total system flow at the pressure drop of the circuit with the greatest resistance. By splitting the loop, the highest-pressure-drop circuit will have only a portion of the tees.

Compare this split-loop piping to the same system with only one loop.

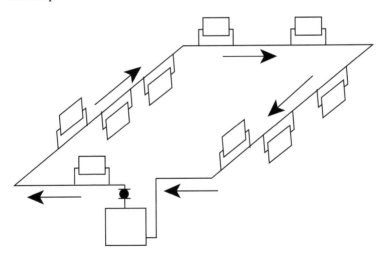

Do you see how all the water has to flow through all the tees now? The circulator has to overcome the cumulative pressure drop of all of those tees. But in the split-loop system, the highest-pressure-drop circuit has only some of the tees. The result is usually a smaller circulator.

Q: How do I size diverter tee systems?

A: The capability of diverter tees varies from manufacturer to manufacturer so it's best to refer to their sizing charts. They'll ask you how much water you have flowing through the main and how much of that water you want to pass into the branch leading toward the radiator. Your BTU/Hr. loads will determine the flow rates in both the main and the branches. If you've sized your system for a 20°F temperature drop from supply to return, each gpm of flow will carry 10,000 BTU/Hr. of heat.

Q: Can you give me an example of how a manufacturer might size a system?

A: Sure. Let's say you have a system with a total heating load of 40,000 BTU/Hr. If you're working with a 20°F temperature drop, you'll want to circulate 4 gpm through the main (each gpm carries 10,000 BTU/Hr. when the temperature drop is 20°F). Now, let's say your first radiator needs 10,000 BTU/Hr. That load represents 1 gpm. So the manufacturer would try to size the diverter tee (or tees) to pass 1 gpm into the radiator. Naturally, 3 gpm would be flowing along the main, between the two tees, while this is going on.

The two flows (1 gpm and 3 gpm) will rejoin on the return side of the radiator to make up the 4 gpm you started with.

Q: Won't this rejoined flow be cooler?

A: Yes, because some of the heat hopped off in the radiator.

Q: How can I tell how much cooler the water will be when it heads off toward the next radiator?

A: If you know the flow rates and the BTU/Hr. loads, you can work a simple formula to find out. It deals with the conditions in the return tee and it goes like this.

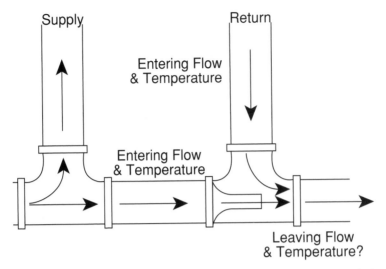

(Entering Flow of the Run) x (Entering Temperature of the Run) + (Entering Flow of the Branch) x (Entering Temperature of the Branch) = (Leaving Flow of the Tee) x ("X" The Unknown Temperature)

Q: Can you give me an example of this?

A: Sure, here are some real numbers.

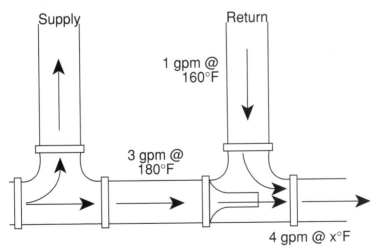

Here we have 3 gpm of 180°F water entering the run of the tee, and 1 gpm of 160°F water entering from the branch. We know the combined leaving flow rate will be 4 gpm. What we don't know is the temperature of that mixed flow.

Plug the numbers into the formula:

$$(3 \text{ gpm}) \ x \ (180°F) + (1 \text{ gpm}) \ x \ (160°F)$$
$$= (4 \text{ gpm}) \ x \ (\text{"X" the unknown temperature})$$
$$540 + 160 = 4X$$
$$700 = 4X$$
$$\frac{700}{4} = X$$
$$175 = X$$

Q: So the temperature of the water heading to the next radiator will be 175°F?

A: That's right. There will be 4 gpm of 175°F water entering that radiator.

Q: Will this radiator have to be larger than the first?

A: It might have to be since it's working with cooler water. It all depends on the heat loss of the space the radiator has to heat.

Q: Does this mean that with diverter-tee systems it's best to supply the areas of largest heat loss first because the water will be hotter?

A: Not necessarily; it depends on the size of the radiators. You're going to have to size each radiator for the space it serves, no matter how you look at it. Where most installers go wrong is by oversizing everything. They're trying to cover themselves, but they usually wind up with over- and underheated rooms and unhappy customers.

Q: Do the manufacturers' sizing charts show all these variables?

A: No, basically they show you the flow rates you can expect in a given situation if you use one or two tees.

Q: Will I always be able to get the exact flow rate I need?

A: Probably not. More often than not you'll have to settle for a little more or a little less. For example, let's say a diverter-tee manufacturer's sizing chart shows you can't get that 1 gpm you were looking for by using one tee because your piping to the radiator is too long. The manufacturer does show, however, that by using two tees, you can get, let's say, 1¼ gpm. It's more than you need, but it will get the job done.

Q: Is there no other way around this?

A: Oh, sure there is. There's usually a way around most everything in hydronic heating, but there's also a price. In this case, you could use larger pipe between the main and the radiator. That would lessen the branch-circuit pressure drop and save you from having to buy that second diverter tee. Whether or not you choose to do this depends on how much time and money you'd have to invest in the decision. Most people, I've found, would opt for the second diverter tee.

Q: But won't that additional diverter tee increase the size of my circulator?

A: It will, and this is one of the things you have to balance when you first look at the system. There's no free lunch!

Q: Why don't we see installers putting in diverter-tee systems anymore?

A: Beats me! I think it's a fine system, but it does require more thought to size than the system we're going to look at next.

CHAPTER FOUR:

LOOP
HOT-WATER
HEATING

"I put a loop of baseboard in this VFW hall and I can't seem to get the end hot. How come?"

"How many feet of element is on the loop?"

"About three-hundred-eighty feet."

"What size baseboard?"

"Three-quarter inch."

"That's much too much element. You're running out of heat."

"Oh, should I use a bigger pump?"

—Conversation during a break at a heating seminar, 1989

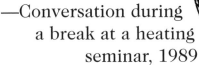

Q: What's loop hot water heating?

A: It's the simplest method of heating with hot water. Each zone consists of a single loop made up of the pipe and the radiators. The water flows out of one radiator into the next.

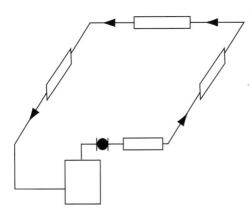

Q: What sort of radiators do most people use with loop hot water heating?

A: Usually fin-tube baseboard. In fact, it was this type of radiation that first made the loop method of heating so popular in the early 1950s.

Q: Why is baseboard so popular?

A: Most heating contractors use baseboard radiation as perimeter heat, running it from room to room along the outside walls of the building. Piped this way, the baseboard radiation becomes the piping as well as the means for transferring heat from the water to the air. Compared to earlier methods of heating, the baseboard loop system is inexpensive and relatively foolproof.

Q: Does this mean I have to use baseboard radiation if I want to install a loop system?

A: Not at all. You can make a loop system with just about *any* type of radiation. All you have to do is pass the water from one radiator to the next in a series.

Q: Is there a drawback to using other types of radiation in a loop system?

A: There can be a drawback to using *any* type of radiation in a loop system, and that includes baseboard radiators. Your success depends on how accurately you sized your radiators to the heat loss of the rooms they're going to serve. If you oversize the first radiators on the loop, the water may be too cool by the time it reaches the last radiators on the loop.

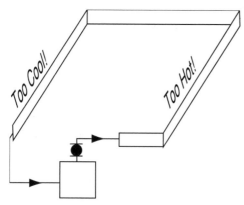

Q: What sort of prob-
lem would that give
me?

A: The last radiators
might not be able to
heat the rooms they
serve on the coldest
days of the year. Your
system would be out
of balance.

Q: How likely am I to
run into this imbal-
ance problem?

A: It all depends on how the builder laid out the rooms and
whether or not the people leave the interior doors open or
closed. Most installers run baseboard from wall to wall. This
looks neat, but it has nothing to do with how much heat the
room needs on a given day. Too much or too little radiation
leads to imbalance and discomfort.

Q: Can you give me an example of this?

A: Sure! Let's say you're putting a baseboard loop system into
someone's house. The first room the loop enters is a bedroom
that measures 15' x 15'. If you run baseboard around the
perimeter, you'll be installing 30 feet of element. Since each linear foot of baseboard puts out about 600 BTU/Hr. (when the
average water temperature is 180°F), your radiator will be
pumping about 18,000 BTU/Hr. into that bedroom.

Suppose the heat loss of that room is only 8,000 BTU/Hr. on
the coldest day of the year? You'll be overheating the room
every time the system comes on.

Q: Won't the thermostat just turn the circulator off if the
room gets too hot?

A: That depends on where the thermostat is. Suppose it's not
in the bedroom. Suppose it's in the living room. Is there enough
baseboard in the living room to shut off the thermostat before
the bedroom overheats? Maybe someone opened the front door
and there's a cool breeze hitting the thermostat. And keep in

mind that since the loop goes to the bedroom before it goes to the living room, the water in the bedroom radiator will be hotter than it is in the living room radiator. That also contributes to the imbalance.

Q: So in a case such as this, would it be smarter to route the loop through the living room first?

A: It depends on whether or not the people living in the house like a cool bedroom. If they do, it would make sense to send the hottest water to the living room first, but remember, there are probably other bedrooms to consider as well on this loop.

Q: How can I solve these imbalance problems?

A: The best way is to size the radiation to the heat loss of the individual rooms. However, if you've already installed the baseboard you can cut down on the amount of heat coming out of each section by closing the dampers.

Q: How does that affect the amount of heat coming out of the radiator?

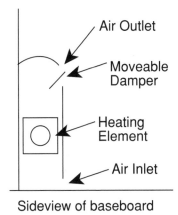

Sideview of baseboard

A: The damper slows the flow of air through the radiator.

Keep in mind this type of radiator works by convection. The air surrounding the radiator picks up the heat from the hot element and rises. Colder air enters the radiator from the bottom to take the place of the rising hot air. If you close the damper, you slow the movement of air and decrease the radiator's BTU/Hr. output.

Q: Suppose I close the dampers and there's still too much heat coming out of the radiator. What can I do then?

A: You could wrap a part of the element in aluminum foil. This will decrease the radiator's surface area and decrease the transfer of heat from the metal to the air.

Q: Could I also remove some of the fins from the baseboard element?

A: Yes, that would work too. By removing the fins, you've decreased the radiator's surface area. Less surface area means less heat transfer.

Q: Suppose I lower the temperature of the water. Wouldn't that also give me less heat in the room?

A: It sure would, and this is a good time to look at the way radiator manufacturers rate their units.

(BTU/Hr. per linear ft. with 65° entering air)

Water Flow	Hot Water Ratings (BTU/Hr.)											
	140°	150°	160°	170°	**180°**	190°	200°	210°	215°	220°	230°	240°
1 gpm	320	380	450	510	580	640	710	770	810	840	910	970
4 gpm	**340**	400	480	540	**610**	680	750	810	860	890	960	1030

Here are the ratings for a popular brand of ¾" copper fin-tube baseboard. As you can see, when the average water temperature of a 4 gpm flow through the baseboard is 180°F, each linear foot of baseboard will put out 610 BTU/Hr. However, if you lower the average water temperature to 140°F, each linear foot of baseboard will only put out 340 BTU/Hr.

Q: When do I need the hotter water?

A: When the outside air temperature drops toward the design temperature. These are the things you consider when you first size the job. You begin with your heat loss calculation. Let's say you want it to be 70°F indoors on a 0°F day. Your heat loss calculation may tell you that a given room will lose 6,100 BTU/Hr. on that day, so you figure the room needs 10 feet of baseboard because each foot puts out 610 BTU/Hr. **when the average water temperature is 180°F.**

On a day when the outdoor temperature is, say, 40°F, you won't have as great a heat loss so you won't need 6,100 BTU/Hr. input. On these days, it pays to run cooler water through the baseboard to prevent overheating.

Q: Do I have to reset the temperature of the boiler water every day?

A: You wouldn't do this yourself, but you could use a "reset" control to do it for you automatically. These controls sense the outdoor-air temperature as well as the boiler temperature and constantly adjust the two to meet the needs of the day. The circulator runs continuously on this type of system.

Q: Will one of these controls solve all my heat-balance problems?

A: They'll help, but they won't solve the problem entirely. You'll still need to match the size of the radiator to the heat loss of the room on the coldest day of the year.

Q: Suppose my baseboard loop serves a large open area. Will I have fewer balance problems in this type of room?

A: Generally, yes. The convective air currents move the heat around the wide open space and distribute the heat more evenly than they could in an area where the builder has partitioned the rooms.

Q: So I could have two loop systems in a house and one could be more comfortable than the other?

A: Exactly. For instance, let's say you have a loop serving the downstairs of a house. The rooms are open to each other, the living room joins the dining room, the family room and the kitchen. Warm air moves freely from one to the other and the people are comfortable.

There's a second loop upstairs in this house, but this one goes from bedroom to bedroom. Since the family members keep the bedroom doors closed at night, some rooms are warmer than others and the people are either too hot or too cold.

Q: I like to run baseboard from wall to wall because I think it looks better that way. How can I avoid overheating problems and still keep those clean-looking lines?

A: You can run the radiator enclosure from wall to wall if you like the way that looks, but you don't have to fill the whole thing with fin-tube. For instance, if you have a 12-foot wall and the heat loss of the room calls for six feet of element, install six feet of element, but make up the difference with bare copper tubing inside the enclosure. This will not only save you some money, it will also increase the level of comfort in the room.

Q: Is there a maximum amount of baseboard element I can

use on a loop?

A: Here again, it depends on how the builder laid out the rooms. If the loop goes through areas where the people are going to close the doors, you have to be very conscious of the average water temperature in the element at the end of the loop. The longer the loop, the greater the temperature drop from one end to the other.

Q: **Can you give me an example of this?**

A: Sure, let's say you're installing ¾" baseboard. If your average water temperature is 180°F, each linear foot of baseboard will put out 610 BTU/Hr. As the water flows, that heat moves into the air, dropping the temperature of the water as it goes. When you get to the end of the loop you won't be getting 610 BTU/Hr. per linear foot anymore. If you haven't sized the baseboard for the lower temperature in that end room you won't be able to get the room to the right temperature on the coldest day of the year.

Q: **What sort of temperature drop do most installers work with?**

A: Usually 20°F.

Q: **How come?**

A: Because with a 20°F temperature drop, the mathematics is easy—each gpm will carry 10,000 BTU/Hr. Also, you leave yourself a margin of safety when you work with a 20°F temperature drop. If you don't have enough heat in a room, you can always raise the boiler temperature a bit to get a higher average water temperature and more heat. The danger of installing too much fin-tube is that the water temperature will drop more than 20°F and be too cold at the end of the loop.

Q: **If my average water temperature is 180°F, what temperature do I start off with?**

A: If you're working with a 20°F temperature drop, you'd start with 190°F at the boiler and end with 170°F at the end of the loop.

Q: **So how much element can I safely use and still stay within the bounds of a 20°F temperature drop?**

A: As a rule of thumb, you should not exceed these limits on any loop:

½"—25 feet of element
¾"—67 feet of element
1"—104 feet of element
1¼"—177 feet of element

Q: Does that include the piping to and from the baseboard radiation?

A: No, it's the active element itself, the part that's open to the air—no closed dampers, no furniture blocking the free movement of air.

Q: Does this mean the system won't work if I exceed these limits?

A: No, it's just a rule of thumb. If you install more element the average water temperature will drop to a point where you may not be able to heat the end rooms to the correct temperature **on the colder days of the year**. On milder days, you probably won't have a problem.

Q: Suppose I need to install 100 feet of ¾" element on a single loop to give me about 61,000 BTU/Hr. How can I do this?

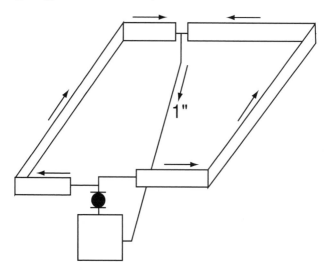

A: The simplest way is to split the loop.

Leave the boiler and head in two directions, assigning part of the 100 foot total to one side and the balance to the other side. Join the two ends together with a single pipe and return to the boiler.

Q: Will that common pipe have to be larger than ¾"?

A: Yes, in this case, it would be 1".

Q: How come?

A: Because it has to carry the combined flow of both sections of baseboard. If the common return pipe is too small, you won't get the flow you need through the baseboard.

Q: What determines the flow I need through the baseboard?

A: The baseboard manufacturer. Let's take another look at that rating chart.

(BTU/Hr. per linear ft. with 65° entering air)

Water Flow	Hot Water Ratings (BTU/Hr.)											
	140°	150°	160°	170°	180°	190°	200°	210°	215°	220°	230°	240°
1 gpm	320	380	450	510	580	640	710	770	810	840	910	970
4 gpm	340	400	480	540	610	680	750	810	860	890	960	1030

Notice how they list the heat output per linear foot at 1 gpm and 4 gpm. This has been a testing standard for many years. The 4 gpm flow rate is a maximum for ¾" pipe because if you make the water move faster than this you'll get velocity noise.

Q: What's that?

A: Velocity noise is the sound water makes when it moves too quickly through a pipe. In hydronic heating, the limits are:

Not faster than 4 feet per second in pipes 2" and smaller.
Not faster than 7 feet per second in pipes 2½" and larger.

Most equipment manufacturers give limits to the velocity they want to see flowing through their equipment. In the case of ¾" baseboard radiation, 4 gpm is the limit.

Q: Can high velocity flow cause any other problems?

A: It can cause erosion of the pipe and early failure of the system. It pays to stay within the limits.

Q: Is that why the common return pipe on the split loop is larger than the baseboard?

A: In part, yes, but that common return also has to handle the combined flow of 8 gpm from the two lengths of baseboard. Remember, you sized that baseboard to deliver 61,000 BTU/Hr. According to the rating chart, you have to circulate 4 gpm through the element to get that per-linear-foot output. That's 4 gpm going each way in the split loop. When the two flows join on the return side, you have to accommodate a total flow of 8 gpm. That's why you need a 1" pipe. One inch can handle the combined flow without velocity noise.

Q: Suppose I joined the two sections of the split loop with ¾" pipe. What would happen then?

A: If the two sides of the split loop were in balance, you'd probably get about 2 gpm flowing through each side. The flow limitations through the common pipe set up what happens in each side of the split loop.

Q: How will this affect my system?

A: You'll get less heat from the baseboard.

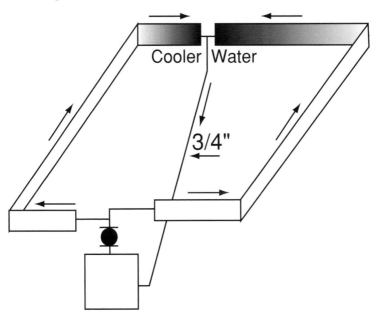

Q: Will I notice this?

A: Probably, but again, only on the colder days of the year.

Q: **What's the best way to get the start-up air out of a split loop?**

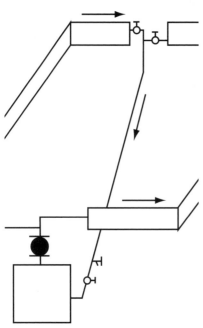

A: Use two shut-off valves, one on each side of the split loop.

Purge the air out of one side and then the other. Make sure you do them separately. If you try to purge both sides through a single valve at the boiler, the air will get stuck in one side and you'll have no heat in that side of the loop. Keep this in mind if you're troubleshooting a no-heat call on a split-loop job. Those shut-off valves are often in the ceiling of a finished basement. You may have to do some work to find them.

Q: **Suppose I'm working with a common, water-lubricated circulator. You know, the kind that come pre-mounted on "packaged" boilers. How long can my total loop be?**

A: Based on the maximum head pressures these little pumps can develop at the flow rates you'd expect to see in a loop system, a good rule of thumb is to keep the total loop (to and from the boiler) under 170 linear feet.

Q: **Suppose my loop has to be longer than that?**

A: You'll have to use a circulator with more head pressure.

Q: **What about a three-piece circulator? They produce less head, so will my loop have to be shorter?**

A: Yes, a good rule of thumb is to keep the total loop length under 130 feet.

Q: **Does the pipe size have anything to do with this?**

A: Not in terms of the pump head, but it does affect the flow rate and the circulator's ability to move heat from the boiler to the radiators. For instance, if you were using a small, water-lubricated circulator on a ½" loop, you could move the water the same distance as you could if you were using a ¾" loop (about 170 feet), but you wouldn't be able to transfer as much heat through the ½" loop as you would through the ¾" loop.

Q: Why does copper fin-tube baseboard sometimes make noise when it gets hot?

A: If you raise the temperature of copper by 125°F (as you will if you start out with 65°F water and end with 190°F water), it will grow by 1.4 inches per 100 feet. That's quite a bit of expansion, and that accounts for the "ticking" noises you frequently hear when the hot water first enters the baseboard.

Q: What can I do about that noise?

A: Many copper fin-tube baseboard manufacturers use plastic gliders to lessen expansion noise. Others offer expansion compensators, which you'd use on long runs to take up the growth of the copper.

Another good way to eliminate expansion noise is to operate the system with an outdoor-air reset control. Set up this way, the circulator runs continuously and the water temperature varies with the outdoor conditions. You don't have the sudden movement of hot water into cold copper as you do with a single-temperature system so you avoid most of the expansion noises.

Q: Every now and then I hear a loud bang in my copper-fin tube loop. How come?

A: It's probably caused by a pipe expanding against a too-small hole in a wooden floor or a wall. Copper grows in diameter as well as in length when heated. If it passes through a hole that's too small, it will "grab" the wood. Then, as it expands in length, it will lift the floor slightly and let it go when there's enough force to break the pipe's grip. That's the bang you hear. You solve the problem by widening the hole (which is easier said than done).

Q: Sometimes I hear a humming sound coming out of the

baseboard. If I tap the enclosure or the element, the noise goes away. What's causing that?

A: Again, if the loop touches something solid such as the floor or a metal beam, it will transmit the sounds of the circulator or the burner through the system. Sound travels further through solids and liquids than it does through air so these vibration noises can show up just about anywhere. The cause and the symptom are sometimes in different rooms. If the noise goes away when you tap the enclosure or the element, look for places where the pipe makes tight contact with the building and give it some space.

Q: If I have to install a baseboard loop on a house without a basement, how can I get past the doors?

A: If the house is on a concrete slab, you'll have to either go over or under the doors. If you go over the doors, the pipe will have to be inside the walls. Be very careful to insulate the pipe well to keep it from freezing in the dead of winter. If you decide to go under the door, you'll have to trench out the concrete.

Q: Can I have problems if I bury the copper pipe in concrete?

A: Yes, because copper and concrete expand at different rates. It might develop leaks as time goes by. Also, some of the ingredients in concrete can be corrosive to copper. In some areas, builders used concrete containing cinder ash, for instance. This really does a job on copper tubing over the years. It's a good idea to insulate the copper from the concrete with a suitable material. Foam pipe covering works well.

Q: Is there a way I can zone each room in a loop system?

A: Yes, you can do it with thermostatic radiator valves.

Q: What are they?

A: A thermostatic radiator valve, or TRV, is a self-contained, non-electric zone valve.

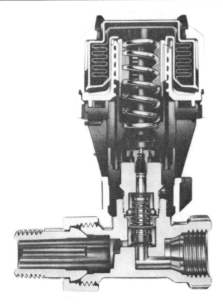

You may remember these from Chapter One. A TRV has two parts: a normally open, spring-loaded valve, and a temperature-sensitive valve operator. You pipe the valve into the line. The operator senses the room temperature and throttles the flow of water through the radiator. You can set the TRV to maintain any room temperature between 50 and 90°F. The circulator operates continuously when you use TRVs.

Q: If I use them on a loop system, won't the first TRV on line shut off the flow to the whole loop when it's satisfied?

A: Normally, it would, but when you use these valves on a loop system, you also use a bypass line around the element.

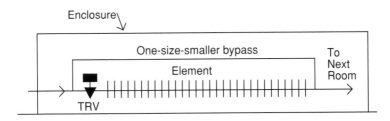

The bypass line is smaller than the baseboard. When the TRV begins to throttle, the water passes over the element and moves on to the next room. In a strict sense, you don't have a one-pipe loop system once you add TRVs, but you do gain a lot of control and solve your heat balancing problems once and for all. TRVs also compensate for heat gain. If it's a sunny day, or if there are a lot of people in the room, the TRV senses the rise in the air temperature and throttles the flow of hot water through the element.

TRVs put the home owner in control.

CHAPTER FIVE:

❧

RADIANT-FLOOR HOT-WATER HEATING

❧

"We may go back two thousand years, only to find that the Romans had a more healthful heating installation than many of those we are making today."

—T. Napier Adlam, 1938

Q: **Do you like this stuff?**

A: You bet!

Q: **Is it difficult to understand?**

A: If I can figure it out, *anybody* can. Now ask me something serious.

Q: **Okay, how's this for starters? What exactly *is* radiant energy?**

A: It's energy that travels through a space in waves (similar to radio waves) and heats only certain things.

Q: **Does radiant energy heat air?**

A: Not to any noticeable degree. Oxygen, hydrogen and nitrogen, the three gases that make up most of what we call "air," can't absorb radiant energy.

Q: **Is there anything in the atmosphere than *can* absorb radiant energy?**

A: Yes, carbon dioxide, water vapor, dust and ozone all absorb radiant energy. This is what concerns scientists who warn us about the depletion of the ozone layer. Without the ozone, too much radiant energy will reach the surface of the earth and lead to an increase in the number of skin cancer cases.

Q: **Are there any other types of radiation that pass through air without heating it?**

A: Yes, microwaves can.

Q: **Is that why the air in a microwave oven stays cool even though the food gets hot?**

A: Yes! The microwaves "see" only the solid objects, not the air.

Q: **Is this what makes a radiant-floor heating system different from a convective heating system such as finned-tube baseboard?**

A: In large part, yes. A radiant-floor heating system first heats the objects in the room. Then, the objects heat the air to a certain degree. They do that by convection, but the movement of the air is relatively slow because the objects in the room don't get that hot.

Q: **What goes on in a convective system?**

A: These systems heat the air *first*, and they heat it to a fairly high temperature.

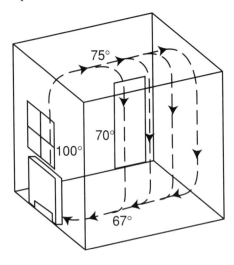

The air then uses its warm, Ferris wheel-like convection currents to heat the people and the objects in the room. In operation, it's the exact opposite of a radiant-floor heating system.

Q: Are there convection currents in a radiantly heated room?

A: None that you notice. You can sense the difference between the two systems right away.

Q: How so?

A: The air in a radiantly heated room is usually very still and it's always cooler than a room heated by convection.

Q: Does this lack of air movement affect the overall level of comfort?

A: Yes. The human body loses about 25% of its heat to convection (drafts) so if the air is still, your body will lose less heat and you'll usually feel more comfortable.

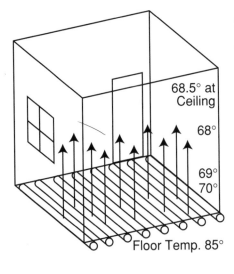

Q: That doesn't make sense. How can I feel more comfortable if the air in the room is cooler?

A: The key difference between a radiant heating system and a convective heating system is that a radiant system strives to control the rate at which your *body* loses heat. A convective system, on the other hand, tries to replace the heat being lost by the *building*. Stop for a minute and think about that because there's a huge difference between those two goals.

Air temperature has little to do with radiant heating. If the air is still and your body is near objects that are as warm as you, you'll feel comfortable, even though the air in the room is cooler than what you're normally used to. The air also feels fresher when it's cooler.

Q: You talk about the rate at which my body loses heat. What's that all about?

A: Your body, believe it or not, is a radiator. At rest, you produce about 400 BTU/Hr. You give up about 100 BTU/Hr. to evaporation when you perspire and breathe. You give up another 100 BTU/Hr. to convective air currents. The rest of the heat, a full 200 BTU/Hr., you give up by radiating toward the colder objects around you.

In a radiant-floor heating system you'll radiate less heat away from your body because the objects around you will be the same temperature as you are. That's what I mean when I say a radiant system *controls the rate at which your body loses heat.*

Q: **Does this explain why I feel uncomfortable when I stand next to a cold surface?**

A: It sure does! Let's say you stand next to a frozen-food case in a supermarket. You feel cold because the warmth of your body moves toward the colder surface of the freezer.

Q: **But isn't the air in the frozen food aisle colder than the air in, say, the cereal aisle?**

A: Not really. The air in both aisles is about the same temperature (check it out the next time you're there). You feel cold in the frozen food aisle because of the enormous difference in temperature between the surface of your body and the surface of the freezer.

Q: **Is there a name for this phenomenon?**

A: For years, heating engineers have called this, "Cold 70." You can be in a room where the air temperature is 70°F, and yet you'll still feel uncomfortably chilly if the surfaces around you are cold because those surfaces will draw too much heat away from your body.

Q: **Is this why I feel cold when I stand next to a single-pane glass window on a cold day?**

A: Yes, even though the air in that room might be 70°F.

Q: **I'm beginning to understand the basic concept. Now, how does a hydronic radiant-floor heating system work?**

A: By circulating warm water through pipes embedded in the floor (or, in some cases, the walls or ceiling) the floor becomes a large "radiator." Radiant waves of energy move off the floor and travel out in all directions to warm the walls, ceiling, and furniture. As these objects become warm, you experience less heat loss because you're standing next to warm things. You become comfortable.

Q: **Does it matter how tall the ceiling in that room is?**

A: No. In fact, in a room heated by a radiant-floor system the air temperature at the ceiling will be nearly the same as it is two feet above the floor.

Q: **That's hard to believe. Can you prove it?**

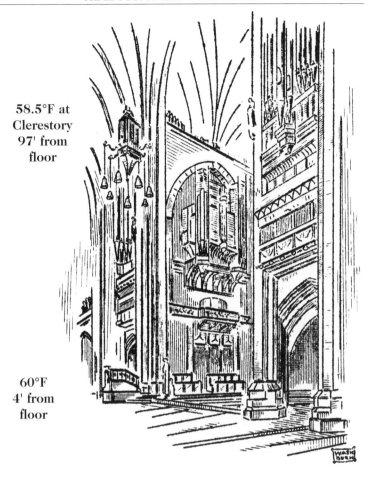

58.5°F at Clerestory 97' from floor

60°F 4' from floor

A: I don't have to; others proved it years ago.

During the 1930s, English heating engineers were developing the science of modern radiant-floor heating. They had a lot of questions and they wanted to see what sort of effect a high ceiling would have on radiant heat so they went to Liverpool Cathedral. This was a large, radiantly heated church with a ceiling that soared over 90 feet.

The engineers learned, to their astonishment, that the air temperature near ceiling (over ninety feet above the floor!) was only 1½°F cooler than the air temperature four feet above the floor.

Q: **How can this be? I've been in tall buildings and the air at the ceiling is always too hot.**

A: I'll bet a convective heating system served every one of those over-heated buildings. Convection currents allow hot air to rise and gather at the ceiling. With radiant-floor heating, your goal is simply to maintain a temperature balance between the people and the objects in the room. You're not directly heating the air so it doesn't rise to the ceiling.

Q: Is the temperature of the floor the same as the temperature of the air?

A: No, the floor will always be warmer than the air in the room. Remember, the floor is the radiator.

Q: Is there an ideal maximum temperature for the floor?

A: Yes, it's 85°F.

Q: Why 85°F?

A: Because that's the surface temperature of most people's clothed bodies. Here, try this. Put a thermometer on your forearm. You'll find the temperature is about 90°F or so.

Now hold the thermometer against your shirt. I'll bet you find it's 85°F.

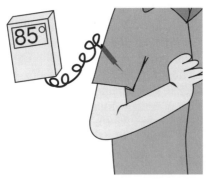

Try it with your friends; we're all about the same temperature (you might even make some new ones with this experiment!). If you're 85°F, and the objects in the room are 85°F, you'll lose hardly any heat by radiation. That's where that terrific radiant-heat comfort comes from.

Q: I thought the temperature of the human body was 98.6°F. Was I wrong?

A: No, you're right about the normal temperature of the human body—on the *inside*. The outside is always cooler than that, and it's remarkably consistent from person to person.

Q: Okay, suppose I decide to put a rug on the floor. Does that mean I have to make it hotter than 85°F?

A: It's the *surface* of the floor that we're concerned with. If there's a rug, you have to make the surface of the *rug* 85°F. That means you'll have to make the floor under the rug hotter than 85°F because the rug and the pad under the rug act as insulators. Rugs on floors are like sweaters on people.

Q: Would I be more comfortable if I made the surface of the floor hotter than 85°F?

A: No, you'd actually be *less* comfortable. If the floor and the objects in the room are hotter than your clothed body, you'll retain too much heat. You'll feel uncomfortably warm, and your feet will perspire.

Q: I'm beginning to see how this heat transfer business works both ways. The goal is to make the surfaces in the room the same temperature as the surface of my clothed body, but not more, right?

A: Exactly. That's what I mean when I say a radiant-floor heating system controls the heat loss of your body. When your body finds its natural heat-transfer balance, you feel wonderfully comfortable.

Q: Is radiant-floor heating new?

A: Not at all! In fact it's probably the oldest method of central heating there is. The Romans used a crude system of radiant heat to warm their famous baths as early as 80 BC.

They built fires and let the heat travel through passages under the marble floors. Europeans heated their castles in much the same way during the Dark Ages.

Squeaky Clean Roman Dudes

Q: **What did they call these systems?**

A: They had a fancy name for them. They called them "hypocausts," which means, "to light a fire under." And that's exactly what they did!

Q: **Did they always use the floor as the radiant surface?**

A: No, in fact in the early 1930s, when the English were pioneering modern radiant-floor heat, they used mostly ceiling radiant panels.

Q: **What's a panel?**

A: It's just an array of tubes buried in the ceiling, floor or walls.

Q: **Did the British install the first panel system?**

A: Actually, the first modern system went into the ceiling of a school in Amsterdam in 1929. The second went into the ceiling of the Bank of England a year later. It was a domed, plaster ceiling with a lot of surface area. They used wrought-iron pipe.

Q: **Why did they use wrought iron?**

A: Because it's tough, malleable and relatively soft. It's easy to work with.

Q: **When and where was the first American radiant heating system installed?**

A: In 1930, in Washington D.C. The British did this one as well. They put coils in the ceiling of their embassy. That system is still working today, although the boiler has been replaced.

Q: **Does a radiant heat system work differently when the coils are in the ceiling?**

A: Theoretically, no. Radiant energy doesn't know up from down. It moves in a straight line, away from its source until it hits something solid. One of the advantages of heating from the ceiling down is that you can often use hotter water. People don't come in contact with the ceiling so you can make the surface temperature hotter than you would a floor.

Q: **Are there drawbacks to heating from the ceiling down?**

A: Yes. Objects such as tables stop and absorb the radiant energy. So, if you're sitting down at the breakfast table, your head and upper body will be warm, but everything below the table will feel somewhat cooler. Also, if you're standing under a low, radiantly heated ceiling, you might begin to feel a bit light-headed.

Q: Does the same thing hold true if the pipes are in the walls?

A: Yes, the side of your body facing the heated wall will always feel warmer than the side turned away from that wall, just as it would if you stood next to a hot oven or an open fire—or a freezer chest in your local supermarket.

Q: Is this the main reason why most of the systems people install today have the tubing in the floor?

A: It sure is! People walk around on the floor, not on the walls or ceiling (hopefully!).

Q: Will a heated ceiling eventually heat the floor?

A: If the radiant energy strikes the floor it will heat it. The heated floor will then heat the air immediately above it.

Q: Does the air at the ceiling get very hot when you're using a radiant ceiling panel?

A: Not at all. A common misconception is that "heat rises." Heat doesn't rise, hot air rises. Heat, well, it radiates! A warm ceiling radiates waves of energy down on whatever surfaces lie below. If that's a floor, the floor gets warm. The same goes for a tabletop, and that's why your lap might feel cooler than your head.

Q: Were there limitations with the early ceiling systems?

A: Because most of their ceilings were made of plaster, the early installers didn't want the temperature of the water flowing through the pipes to be hotter than about 85°F. If the water was too hot, they reasoned, the pipes would expand and contract too much. They didn't want the ceiling to come crashing down on the occupants, so they kept the temperature low.

Q: Can you heat the surface of the ceiling to 85°F using 85°F water?

A: You can heat the ceiling *almost* to 85°F, but it takes a long time.

Q: Did they ever use hotter temperatures for those ceiling installations?

A: As their experience grew, installers realized the plaster ceiling and the embedded metal pipes expanded at about the same rate. They made provisions for the expansion of the *entire* ceiling by leaving a space at the joints (concealed by molding). In time, they began to circulate water as hot as 165°F through the ceiling pipes with no problem at all.

Q: How about when the tubing is in the floor? What determines the temperature of the water you need inside the tubing?

A: The temperature depends on three things: how close together and how deeply buried the tubing is, what the floor is made of, and how well constructed the building is.

The typical water temperature for radiant floor heating will vary between 90 and 160°F, depending, in large part, on whether the tubing is in a slab or stapled under a wood floor, and whether or not the floor is covered with a rug or other insulating material.

In general, the more the flooring material surrounds the tubing, the lower the water temperature can be. In other words, if you bury the tubing in concrete there will be more tubing-to-floor contact than there will be if you staple the tubing to the underside of the wood floor. The result is the water temperature can be lower when the tubing is in concrete.

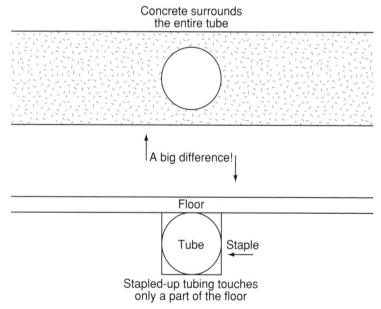

Concrete surrounds the entire tube

A big difference!

Floor

Tube | Staple

Stapled-up tubing touches only a part of the floor

Q: So the heat is not really moving out of the tubing by radiation, is it?

A: No, the heat is moving out of the water, through the walls of the tubing, and into and through the flooring material by conduction. The radiation aspect comes in only when the heat reaches the surface of the floor and leaps off.

Q: What's conduction?

A: It's the type of heat transfer you get when a hot object touches a cooler object. Try holding a length of copper tubing while you solder a fitting onto its end. Ahhhhhgggggg! That's conduction.

Q: Got it. Now, we don't want the floor's surface temperature to exceed 85°F, no matter what the temperature of the water is, right?

A: That's right. If the floor is too warm, the people in the room will be uncomfortable. Don't confuse the temperature of the water with the temperature of the floor's surface. We heat one (the water) to achieve the other (the surface temperature of the floor).

Q: Can a warm floor damage the furniture by drying it out?

A: During the 1940s, manufacturers involved in the radiant floor heating industry did extensive testing on this issue. They found that as long as air could circulate around the furniture, the temperature of the furniture would not exceed 85°F. This was not hot enough to cause any sort of drying-out damage. In fact, Bell & Gossett published findings at the time stating that radiant-floor heating was less damaging to furniture and draperies than "conventional" types of heating systems.

Q: Can a radiant-floor heating system damage rugs?

A: Again, research done by manufacturers during the 1940s showed that rugs were unaffected as long as the maximum surface temperature of the floor was 85°F or less. The same applies to varnish or wax on a wood floor. Eighty-five degrees F surface temperature is similar to summer sunlight on the floor.

Q: Who first put pipes in the floor? Was it also the British?

A: No, it was the Americans. They did it right after World War II. Before the war, heating engineers were reluctant to bury metal pipes in concrete because of concerns about corrosion to the pipe. However, after the war, European and American engineers had the opportunity to study the metal reinforcing bars in bombed-out European buildings. They found them to be in remarkably good shape. Inspired by these studies, American heating engineers began to use metal tubing in tract housing. They buried the tubing right in the concrete and found this to be a fast and inexpensive way to get the job done. By the end of the 1940s, 95% of American radiant heating systems were floor-heating systems. The balance were ceiling-heating jobs with a few wall-heating systems.

Q: Why did installers avoid putting pipes in the walls?

A: Because people drive nails into walls to hang pictures. Also, if you're installing the pipes on an outside wall, you have to use more insulation than normal to keep the heat from moving toward the outside instead of the inside.

Q: Couldn't I install the pipes within an inside wall?

A: You could, and that would get you around the insulation problem, but you'd also set up warm and cold "zones" within the house, which you'd notice when you moved around the rooms.

Q: Was there an economic advantage to floor heating over ceiling heating?

A: Ceiling-heating systems used to cost 10 to 20 percent more than floor-heating systems.

Q: What company was at the forefront of radiant-floor heating for homes?

A: Chase Brass and Copper Company. They installed a radiant heating system in the ceiling and the floor of a New England house in 1942.

They used the house as a laboratory during the winter of 1942-43 and performed extensive tests. In 1945, they published their results in their classic *Chase Radiant Heating Manual* and set the stage for the post-World War II building boom.

Q: What type of pipes did American builders use during that building boom?

A: Copper, steel and wrought iron. Of these, installers and manufacturers of the time considered wrought iron to be the most durable.

Q: Were there problems with these systems?

A: With many of them, yes. But most of these problems were caused by the materials and the "Get it done fast!" work habits of the time. Consider, for example, William Levitt who built 50,000 mass-production "Levittown" homes in New York and Pennsylvania at the astonishing rate of one every two hours. Every one of them had a radiant-floor heating system.

Q: What sort of problems developed in these systems?

A: Most of the Levittown problems were a result of the hastily installed concrete slabs that shifted and cracked over the years, causing the metal pipes to stress and leak.

Q: Is concrete itself corrosive to metal pipes?

A: It can be. It all depends on what goes into the concrete. For instance, some concrete contains cinders from incinerators (that's where the name "cinder block" comes from). Cinders contain sulphur compounds that can damage metal pipe.

Q: So what material should surround metal pipe when you bury it in the floor?

A: The old-timers used broken limestone, gravel, sand or non-sulphur-bearing concrete. If their connecting lines to and from the boiler ran through cinder fill, they dug a trench and filled it with crushed limestone. They covered the trench with a slush coat of lime mortar. The limestone neutralized the acids in the cinders and prolonged the life of the pipe.

Q: What basic installation rules were they supposed to follow in those days?

A: There were several, according to the "how to" books of the time:

1. The edge insulation should extend vertically, three feet below the slab (or at least below the frost line).
2. The copper tubing should never come in contact with cinders

or other corrosive fill.

3. The installer should handle the tubing with care so it doesn't get dented.

4. The installer should test the tubing under 200 psig for at least 4 hours.

5. The installer should never step on the tubing or run a wheelbarrow over it.

6. The installer should be especially careful when spading or hoeing the fresh concrete over the tubing.

Q: Did the tract-housing builders follow these rules?

A: Most of the time, no. There simply wasn't enough time.

Q: Is the radiant-floor tubing that's available today better than copper, steel and wrought-iron?

A: Yes, much better. Today, there is plastic and rubber radiant-floor tubing that can take relatively high temperature and pressure. These materials won't corrode in the presence of concrete, ground water or dissimilar metals.

Q: Is polyethylene one of these plastics?

A: Yes, one of the oldest, in fact. Researchers discovered polyethylene in the 1930s completely by accident.

Q: How did that happen?

A: In 1939, a group of English chemists were studying the reaction between ethylene and benzaldehyde. They were cooking this brew in an autoclave under extremely high pressure. Suddenly, the autoclave exploded and destroyed the entire laboratory. Fortunately, it didn't hurt anyone.

When the scientists crept back into the lab and looked into what was left of the autoclave, they found a white, waxy substance. No one knew what this stuff was at first, but further study of the molecular structure revealed it was a gas that the extremely high pressure within the autoclave had turned into a solid. It was the first "plastic."

They called it polyethylene and then, for the next several years, tried to make it again. Finally, out of frustration, they duplicated their original experiment (by blowing up another laboratory!).

It didn't take them long to realize that oxygen, introduced at high pressure, was the catalyst they needed to make polyethylene. The rest is history.

Q: Did they use polyethylene to make radiant-floor tubing?

A: No, that came later. At first, they used it in the development of radar and the undersea cable. They found it to be very flexible and not subject to corrosion. They used it to make many things.

Q: Did it have any drawbacks?

A: Yes, its main drawback is that it gets soft when heated to higher temperatures. However, within the low-temperature range of most radiant heating systems, polyethylene works well.

Q: What if I use it on a radiant-floor system and the boiler temperature should go too high?

A: Polyethylene radiant-floor tubing will soften and eventually fail if the water gets too hot for too long a time.

Q: How can I protect the tubing from high temperature?

A: You can use additional high-limit temperature controls or lockout solenoid valves to keep water that's too hot from moving into the tubing.

Q: When did they first begin to use polyethylene for radiant-floor heating?

A: In the 1950s.

Q: How was it received by the heating community?

A: With a lot of reservation. Most of the concerns at the time revolved around the effects of high temperature and chemicals such as antifreeze. Another problem in the early days was there were a number of fly-by-night companies producing plastic tubing from scrap material.

Q: Weren't there any testing procedures or standards for the tubing?

A: Not in those days.

Q: What about this material called "PEX?" Is this also polyethylene?

A: No, PEX starts as a material known as high-density polyethylene (PE-HD), and is then changed by one of several processes.

Q: What does "cross-linked" mean?

A: It means the individual polyethylene molecules have joined on their carbon atoms. PE-HD is a cross-linked material, but PEX differs because its chemical bond is three-dimensional. This gives the material a thermal "memory," meaning that if you kink it, you can warm it and it will return to its original shape.

Q: What does cross-linking do for the material?

A: It turns it into a "macro-molecule" (remember, the molecules linked together on their carbon atoms). This gives the material additional strength, allowing it to take more mechanical stress.

Q: How do manufacturers make PEX?

A: There are several chemical methods, and also a method that uses irradiation to link the molecules.

Q: Is one method better than the other?

A: The chemical methods cross-link the material while it's in a liquid (melted) stage. This results in a more uniform distribution of the cross-lining sites. The irradiation method, on the other hand, treats the material while it's in the solid state. The cross-linking results are less uniform with the irradiation method.

Q: Wait a minute. Aren't they supposed to cross-link all of the material?

A: No. If there are too many cross-lined sites the material will become brittle. If there aren't enough, the material will be no better that the material it started out as (PE-HD).

Q: Who invented PEX?

A: Thomas Engel, a European scientist, first created it in the late 1960s. The Swedish pipe manufacturer, Wirsbo, first marketed it as radiant-floor heat tubing in the early 1970s.

Q: What is polybutylene?

A: Polybutylene (commonly known as PB) is a flexible plastic that has been used for radiant-floor heating systems for over 25 years. It's not a "cross-link" material, but it stands up very well to the normal conditions found in a radiant-floor heating system. Many installers prefer polybutylene because of its relatively low cost.

Q: Is polybutylene more flexible than PEX?

A: Yes. Tests show that at 95°C (203°F) polybutylene has a better resistance to pressure than PEX. This means that polybutylene can have a thinner wall than PEX, giving it greater flexibility while decreasing its overall weight.

Q: What is EPDM?

A: EPDM is an acronym for (are you ready?) ethylene propylene diene monomer. Essentially, it's a type of rubber that has been commonly used in many plumbing and heating products for years. Several manufacturers make a radiant-floor hose out of EPDM.

Q: Are there other rubber materials besides EPDM?

A: Yes, some manufacturers are now making radiant-floor hose by layering several types of industrial-grade rubbers together.

Q: Is one radiant-floor tubing material better than another?

A: It depends on who you listen to. Naturally, each manufacturer preaches the virtues of his material over his competitors', but from what I've seen and heard, all the materials being supplied by reputable manufacturers work very well—providing you follow their installation instructions. I've heard of no major problems with any of them.

Q: Which material is the most flexible?

A: The rubber-based hose is generally more flexible than any of the plastics.

Q: How tightly can I bend these materials?

A: It depends, in part, on the size of the tubing. For the common radiant-floor sizes, most plastic tubing manufacturers recommend you not make a bend tighter than six inches. Rubber-based hose manufacturers allow you to make a bend on a 4" radius.

Q: How does the air temperature affect these materials?

A: The plastics get rigid when the weather gets cold. Most plastic tubing manufacturers say you should try to install their tubing when the outdoor temperature is above 50°F. If you can't, they recommend you keep the tubing warm until you're ready to use it.

The rubber-based hose remains pliable at temperatures well below freezing.

Q: **Is radiant heat popular in Europe?**

A: Yes, they use radiant-floor heating in about 25% of their new residential construction. And keep in mind, in Europe, literally *every* building has hydronic heat.

Q: **What radiant-floor tubing material do the Europeans use?**

A: Nowadays, it's mostly PB (polybutylene) and PEX (cross-linked polyethylene), but they also use other types of plastic tubing as well. The Europeans have not been using rubber radiant-floor hose.

Q: **Why don't the Europeans use rubber radiant-floor hose?**

A: Because their more-mature market developed around the plastics: polyethylene, polybutylene and PEX. Rubber radiant-floor tubing is essentially an American development that came out of the solar industry during the 1970s.

Q: **How long have the Europeans been working with radiant-floor heating?**

A: Since around 1972.

Q: **Have they had any growing pains?**

A: Yes, the greatest being the problem of oxygen diffusion through the radiant-floor tubing.

Q: **What's oxygen diffusion?**

A: It's the movement of oxygen through the walls of the radiant-floor tubing.

Q: **How is this possible?**

A: By a phenomenon called osmosis. Unlike metal, plastic is permeable to gasses. It's hard to believe, but even though the pipe is buried in the floor and under pressure, oxygen can *still* move through the wall of the radiant-floor tubing and get into the water.

Q: **What's osmosis?**

A: It's the movement of a gas or liquid through a semi-permeable membrane from an area of higher concentration to an area of lower concentration. In this case, there's more oxygen in the air than there is in the system water. Since the radiant-floor tubing is a semi-permeable membrane, the oxygen moves from the air into the water.

Q: What's a semi-permeable membrane?

A: It's a material through which some, but not all, molecules will pass. Plastic and rubber are both semi-permeable. Metal is not.

Q: How does the oxygen get through the concrete?

A: Concrete isn't air tight.

Q: But if the water in the system is under pressure, how can anything get in from the outside?

A: When we consider osmosis, we're not concerned with the pressure of the system. We're concerned with the "pressure" of the molecules, in the sense of "How many are there?". In other words, what matters is the *percentage* of a certain gas on one side of the semi-permeable membrane compared to the *percentage* on the other side of the membrane. Nature will always try to balance the two percentages to find equality. That's why the oxygen moves through the wall of the radiant-floor tubing and into the system water.

Q: Is oxygen bad for the radiant-floor tubing?

A: Not at all. Oxygen can't damage the plastic or rubber radiant-floor tubing, but if enough oxygen gets into the water, it *will* attack the ferrous parts of the system and create a sludge that can eventually plug up the radiant-floor tubing, control valves and the circulators.

Q: When you say the "ferrous" parts of the system, what do you mean?

A: Any metals that rust in the presence of oxygen are "ferrous" metals. We call metals such as brass and bronze "non-ferrous" because they don't rust. Iron and steel are both "ferrous" materials.

Q: How did the Europeans overcome this oxygen diffusion problem?

A: They did it in several ways:

1. They started to treat the water with a corrosion inhibitor. (The main drawback of this method is that someone has to monitor the concentration of inhibitor from year to year).

2. They began to separate the radiant-floor tubing side of the system from the ferrous metal side of the system by using a heat exchanger. (You can also use a copper finned-tube boiler

and non-ferrous circulators and valves to achieve the same results without a heat exchanger).

3. They developed and began to use barrier radiant-floor tubing.

Q: What's barrier radiant-floor tubing?

A: This is special radiant-floor tubing that has a protective sheathing either on the outside or the inside of the pipe. The technical name for this sheathing is an EVOH.

Most plastic tubing manufacturers apply the barrier sheathing to the outside of their radiant-floor tubing. That's what gives the tubing its shiny appearance. One rubber radiant-floor hose manufacturer places the oxygen barrier between two layers of rubber on the inside of the radiant-floor hose.

Q: Is this type of EVOH barrier material used only in radiant-floor hose?

A: No, it's widely used throughout many industries. The food processing industry, for example, uses EVOH barrier material to extend the shelf life of things such as potato chips (those shiny bags are made from an EVOH). Mylar balloons are made from an EVOH. That's why they hold helium for such a long time. The chemical industry uses EVOH liners in workers' clothing to keep harmful gasses from coming in contact with the workers' skin.

Q: Is barrier radiant-floor tubing expensive?

A: Well, it's more expensive than non-barrier radiant-floor tubing, but when properly installed, it's very effective in stopping oxygen diffusion problems.

Q: Can anything cause a problem with barrier radiant-floor tubing?

A: If the water temperature exceeds the manufacturer's set limit, and if the EVOH sheathing is on the outside of the tubing, it can peel off like a snake skin. Manufacturers call this "delaminating."

Q: How can I protect myself against this possibility?

A: In the same way you would if you were using polyethylene or polybutylene tubing: by using additional high-limit temperature controls and lock-out solenoid valves to keep the too-hot water from flowing into the tubing.

Q: Is there anything else I have to watch out for if the EVOH barrier is on the outside of the tubing?

A: Yes, be careful not to scratch the barrier when you install the radiant-floor tubing. If the barrier is damaged, oxygen might work its way into the system.

Q: Is the oxygen-diffusion problem magnified by certain conditions?

A: Yes, mainly by water temperature. The hotter the water, the greater the potential for oxygen-related problems with *any* radiant-floor tubing. Usually, there are few problems if you keep the water temperature below 140°F.

Q: What are the typical temperatures for a radiant-floor system?

A: Usually, in-slab designs call for a maximum of 130°F supply-water temperature. I rarely see oxygen-related problems on these jobs. You have to pay closer attention to "staple-up" jobs where you're attaching the radiant-floor tubing directly to the underside of a wood floor. Since the floor comes in contact with only a small part of the tubing, you have to run higher design temperatures (as high as 160–170°F) on these jobs. The potential for oxygen-diffusion-related problems is much greater here if you're not using either barrier pipe, chemical corrosion inhibitors or a non-ferrous heat exchanger.

Q: What is the "DIN" standard for radiant-floor tubing?

A: This is a German-government standard accepted by the European and American hydronics communities for radiant-floor tubing. They call it DIN-4726 and it sets what the experts consider a "safe" limit on how much oxygen can enter a radiant-floor heating system. DIN-4726 allows for the entry of 0.1 milligram of oxygen, per liter of water, per day when the water is 40°C (104°F). That's an incredibly small amount of oxygen.

Q: How small is it?

A: To give you an idea, let's say you have a 200,000 BTU/Hr. radiant-floor system that contains about 70 gallons of water. To meet DIN-4726, you'd have to make sure not more than *0.0009 ounces of oxygen, per day* entered that system. Not much, is it?

Q: Do all manufacturers now meet DIN 4726 for barrier tubing?

A: Most tubing manufacturers now have an offering that meets or exceeds the DIN 4726 standard. This tubing usually costs a bit more than other types of tubing.

Q: Does this mean I have to use DIN 4726-standard tubing on all my jobs?

A: Not at all! The standard is just that—a *standard*, which you can meet in one of several ways: by adding a corrosion inhibitor to the system water, by separating the ferrous from the non-ferrous parts of the system with a heat exchanger, or by using the barrier tubing. It's your choice.

And again, keep in mind you'll rarely see oxygen-diffusion related problems when you keep the system water temperature below 140°F.

Q: Does protecting the system from oxygen by any of these three methods mean I don't have to use an air separator?

A: No. A properly designed hydronic heating system needs an air separator to remove dissolved oxygen from the water as the water is heated. Ideally, you should pipe the air separator near the boiler where the water is hottest.

Q: Is there anything else I need to watch out for with radiant-floor tubing?

A: Yes, if you're using PEX radiant-floor tubing, make sure you don't expose it to ultraviolet light for too long a time. Sunlight accelerates the aging of this type of radiant-floor tubing.

PEX manufacturers aren't specific in saying how long the tubing can be out in the sun. They refer only to "extended periods of time"; however, from what I've heard, it shouldn't be out in the sun for more than seven days.

Q: Can sunlight damage rubber radiant-floor hose?

A: Not as easily, if at all. One manufacturer, in fact, warrants one of their rubber-based products for continuous exposure to sunlight. They do note that the color of the tubing will fade when you leave it out in the sun, but this has no effect on the product's performance.

Q: Are these systems difficult to install?

A: Not usually, especially with the accessory equipment and support most reputable radiant-floor system manufacturers offer nowadays.

Q: What's the best way to install a radiant-floor heating system?

A: These systems work best when you can bury the tubes in the flooring material.

Q: Why is that?

A: Because there's better heat transfer when the entire outside surface of the radiant-floor tubing touches the flooring material.

Q: What size tubing goes into most residential systems?

A: It's usually ⅜" or ½".

Q: Is it absolutely necessary for the tubing to go into the flooring material?

A: No, you can also install the tubing between the floor joists of an existing building. However, on these jobs, only a thin edge of the tubing touches the flooring material, so you have to use more tubing, and hotter water.

Q: What's involved in a concrete installation?

A: On slab-on-grade jobs, you'll start with a layer of insulation. On top of the insulation, you'll have a plastic vapor barrier. Over that, you'll have steel, reinforcing mesh.

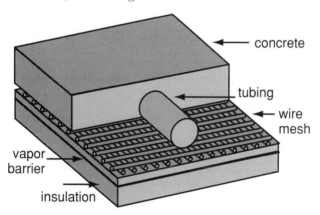

Attach the radiant-floor tubing directly to the mesh, using either cable ties (if you're using rubber hose) or plastic clips (if you're using plastic tubing). Don't use any kind of adhesive tape on PEX tubing as the adhesive may accelerate the aging of the tubing. Make sure you follow the manufacturer's specifications as to tube spacing.

Q: Is it important to insulate the underside of the slab?

A: Well, it makes no sense to heat the ground under the slab. You want the heat to travel upward into the room. Insulation becomes especially important if the people are using rugs or other floor coverings. The thicker the floor covering, the hotter you have to make the slab to heat the rooms. So why waste heat by pushing it downward?

Q: What's the normal temperature of the ground under a house?

A: It averages 50°F under a house, and while earth itself is a good insulator, you'll have to expend energy bringing it up to temperature. Below-slab insulation reduces operating costs and really pays in the long run.

Q: Do I have to insulate under the *entire* slab?

A: It's best if you do, but if you can't, insulate at least under the first four feet in from the edge of the slab.

Q: How about if the room is on the second floor? Do I still have to insulate under it?

A: Yes, and here you *have to* insulate under the entire floor. The insulation keeps the heat from traveling down toward the first floor. If you don't insulate, you could wind up with a situation where the first floor is too hot while the second floor is cold.

Q: How thick should the insulation be?

A: If you have a heated area under the radiant floor, use a minimum of 3½", foil-faced, batt insulation. If the space under the radiant floor is unheated (on a staple-up job), use a minimum of 5½", foil-faced batt insulation.

Q: Should I insulate around the perimeter as well?

A: Yes, this is the most-important place to insulate. Ideally, you should have an unbroken line of edge insulation extending vertically down three feet below grade.

Q: What does that do for me?

A: It keeps the heat from migrating out into the soil surrounding the slab. Without that edge insulation, it will be very difficult to heat the building as promised. This is especially true if the folks are using carpets. A carpeted slab floor can lose as much as 50% more heat through the perimeter than a bare slab floor.

Q: Is there a way to figure how much heat I'll lose if I don't insulate the perimeter edge?

A: Each perimeter foot of slab will lose about 2 BTU/Hr. per degree F difference between the outside air temperature and the slab temperature.

Q: Can you give me an example of this?

A: Sure, let's say you have an uninsulated slab at 85°F on a day when it's 20°F outside. The difference between 85 and 20 is 65. Multiply 65 by 2 BTU/Hr. and you'll see you're losing 130 BTU/Hr. through every linear foot along the slab's edge.

Now suppose the slab measures 50' x 35'. That means you'll have 170 linear feet of uninsulated edge to deal with. If each foot loses 135 BTU/Hr. on that 20°F day, you'll be wasting 22,950 BTU/Hr. (But then again, you'll probably be able to grow tulips in January!).

Q: What type of insulation should I use under and at the edge of a slab?

A: Sheets of polystyrene work well.

Q: What does the vapor barrier do?

A: It keeps ground moisture from working its way up into the slab and, ultimately, into the building. Ground water creates more than just a humidity problem in the building. Left unchecked, it can carry off a great deal of the slab's heat.

Should any leaks develop as the years go by, the vapor barrier will help you spot them. Without the vapor barrier, the water will leak down into the ground, but with the vapor barrier, the water will rise up through the concrete where you can see it.

The problem with many of the radiant-floor-heated homes built during the post-World War II years was that the installers didn't use a vapor barrier under the slab. By the time they discovered the leaks, it was too late.

Q: Is there a way to tell if I have leaks before I bury the tubing in concrete?

A: You have to put the tubing under water- or air-pressure (water is much better) while you're pouring the concrete. This is part of all tubing manufacturers' standard installation procedure.

Q: How much pressure should I use to test the tubing?

A: Between 50 psig and 100 psig.

Q: **How deep in the slab should I bury the tubing?**

A: For radiant-floor heating, the top of the tubing should be about ¾" below the surface, although it can be set as deep as 3 inches. Keep in mind, though, that the deeper you bury the tubing, the longer it will take for the radiant floor to respond to temperature changes in the room, and the hotter the circulating water will have to be.

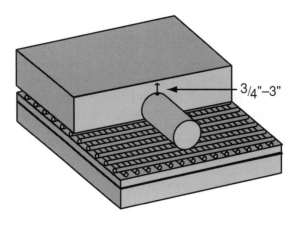

Be careful when you're working within the same zone not to bury the tubing at different depths. If you do, the zone will feel cool in some places and warmer in others. Keep the depth of the tubing as uniform as possible.

Q: **Suppose I'm pouring light-weight concrete over a wooden floor. How do I attach the tubing on these jobs?**

A: Usually, you'll just staple the tubing to the sub-floor, making sure you follow the manufacturers specifications on tube spacing.

Q: **Should I insulate under the subfloor on these light-weight concrete jobs?**

A: Yes, you should, with foil-face insulation, making sure the foil faces up toward the floor.

Q: **Why is that important?**

A: Because the foil reflects the radiant waves of energy back up into the floor. Without it, you'll lose more than 50% of the heat output of the tubing.

Also, leave a 2" air gap between the foil and the underside of the floor. This rule also applies if you're installing the tubing in mud-set under a tile floor.

Q: What's the reason for the 2" air gap?

A: It helps the heat diffuse more evenly between the tubes. The waves of radiant energy sail through the air gap and reflect up after bouncing off the reflective foil. This helps limit the phenomenon called "striping" where you feel most of the heat directly over the tube.

Q: Is there a better way to avoid "striping?"

A: You can use metal heat diffusion plates.

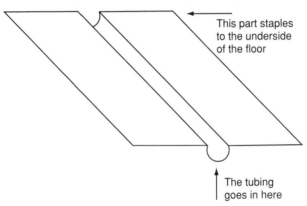

This part staples to the underside of the floor

The tubing goes in here

These plates use conduction to move more of the heat out of the tube and into the floor.

Q: Is there any drawback to the heat diffusion plates?

A: They can be difficult to install if there are nail tips sticking through the floor. The plates have to come in complete contact with the floor to be effective. That means you may have to spend a lot of time snipping nail tips before you can go to work installing the tubing.

Q: Can I use diffusion plates in ceilings and walls?

A: Yes, and they usually make the installation easier.

Q: Do I have to snap the tubing into the plate first?

A: One manufacturer, Radiant Technology, Inc., makes a product called RADIANT-TRAK. These plates get installed first. The tubing snaps in afterwards. These are very easy to work with.

Q: Is there any other way I can attach the tubing to a wood floor?

A: You can install it between "sleepers" on a subfloor.

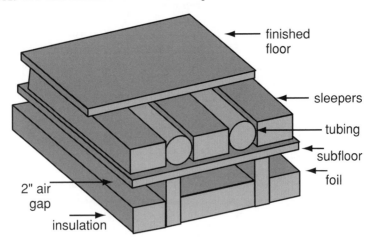

finished
floor

sleepers

tubing

subfloor

foil

2" air
gap

insulation

Just make sure the tubing touches the underside of the finished floor. Contact between the tubing and the floor is critical if you're to pass BTUs from the water into the room. Oh, and watch where you hammer those nails when you put down the floor's top boards.

Q: Suppose I want to attach the tubing to the underside of a wooden floor on the second floor of a house. Can I do this without ripping open the first floor's ceiling?

A: Yes, you can. Let's say the people are putting up a dormer. They'll be exposing the top of the joists in the first-floor ceiling. What you have to do is install 3½", foil-faced batt insulation in that exposed area between the joists, making sure the foil faces up.

Next, stretch chicken wire over the entire floor and set your tubing on top of the wire. (You'll move from joist bay to joist by passing through a series of holes you drilled through the joists.) Now when you install the finished floor, the chicken wire will hold the tubing firmly against the underside of the wood.

Here's a picture of what I'm talking about.

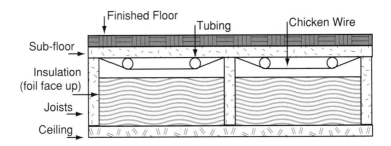

Q: **Is there any way of knowing where the tubing is after I've installed the floor over the tubing?**

A: One of the best ways is to videotape the tubing before you finish the floor. That gives you a permanent record of the tubing layout. You can also use an infrared sensor to see where the tubes are.

One manufacturer, Heatway, makes a tubing that has a radio transmitting wire built into it. You can use a special receiver to "see" the buried tubing after its installed.

The old-timers used cats to find the leaks in the copper tubing. When it wants to lie down, a cat will always look for the warmest spot on the floor. The old-timers would lower the shades in a house to keep out the sunlight and then let the cat loose. It never took more than a few minutes for the cat to find that leak.

Pretty creative, eh?

Q: **How many people does it take to put a radiant-floor heating system in?**

A: That depends on the job and the amount of experience the people putting the stuff in have had. You'd better figure on at least two people, though. You'll also need a tubing uncoiler to help you unwind the tubing. Once you get the hang of it, it's not difficult to install the tubing. It is, however, more involved than a simple baseboard loop system.

Q: **Is it more expensive to install than a baseboard system?**

A: Yes. Most contractors find that when they compare the costs of material and labor between the two methods, radiant-floor heating costs about 25% to 30% more than a copper finned-tube baseboard job.

Q: **Is it less expensive to operate?**

A: It depends on what you compare it to. Over the years, I've heard claims of from 25% to 40% (the greater number coming from a comparison of radiant-floor heating to "scorched air" furnace heating).

I wouldn't hesitate to tell a home owner that his new radiant-floor heating system will provide unparalleled comfort while using *at least* 15% less fuel.

Q: Do you have to be an engineer to size a radiant-floor heating system?

A: Not at all! Most reputable manufacturers offer great technical assistance and easy-to-use sizing charts. Most are now also offering user-friendly computer software to make the sizing a snap.

Q: Is the heat-loss calculation for a radiant heat system different from that of a convective heating system?

A: Yes, and one of the reasons for this is because, in a radiant-floor heating system, the temperature of the air at the ceiling is nearly the same as the temperature of the air near the floor.

Q: Why should that make a difference to the heat loss of the building?

A: Because the difference in temperature between the inside and outside directly affects the speed at which heat moves out of a building. The warmer it is inside, the faster heat will leave the building. In other words, heat goes to cold.

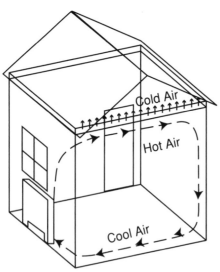

When the temperature at the ceiling is abnormally high, the heat transfers to the Great Outdoors much more quickly. This is typically what happens with "scorched air" furnace heat and, to a lesser degree, with convective hydronics. It's the reason why you have to put more insulation in the ceiling than you do in the walls.

In a radiant system, the movement of heat through the ceiling is less because the air temperature at the ceiling is lower. As a result, the overall heat loss from the building is less.

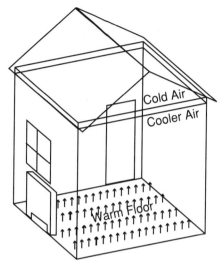

Q: What else makes a radiant heat-loss calculation different from a convective heat loss calculation?

A: Since there's very little movement of air in the room (because there are no convectors or blowers to set up convective currents), there's less infiltration of colder, outside air into the room. That convective movement of warm air increases heat loss to the outside by zipping past the cooler windows and doors. Get rid of the movement of air past the windows and doors and you'll lessen the infiltration heat losses from the house.

Q: Can I use the I=B=R, ASHRAE or ACCA method to calculate a radiant heat loss?

A: You can, but it will most likely give you an oversized system if you take the load calculations at face value.

Q: Is there a rule of thumb I can use to adapt these well-known heat-loss methods to a radiant-floor heating system?

A: First, I want to urge you to use the heat-loss methods developed by the radiant-floor heating industry, if you possibly can—particularly the new computerized programs. However, if you don't have access to this information, you can easily knock 15% off your I=B=R, ASHRAE or ACCA calculation as a rule of thumb. Try it once and you'll see what I mean.

Q: Before I commit to the job, is there a way I can get an idea of whether or not I can heat the room solely with radiant-floor tubing?

A: Yes, there is. Measure the length and width of the room and multiple the two together to get square footage. Now, multiple the square footage of the room by 35 BTU/Hr., per square foot. This will give you the total *radiant* heating capability of the floor in BTU/Hr.

Q: Can you give me an example of this?

A: Sure. Let's say you have a room that's 25 feet wide and 20 feet long. Multiply those two numbers together to get 500 square feet. Now multiply 500 square feet by 35 BTU/Hr./Sq. Ft. (the maximum BTU/Hr. output you should allow for each square foot of floor area) and you'll get the total radiant heating capability of this floor: 17,500 BTU/Hr.

Q: Why are you limiting the floor to 35 BTU/Hr. per square foot?

A: Because 35 BTU/Hr. will bring the surface of the floor to 85°F. This, you'll remember, is the surface temperature of the clothed human body.

It's a classic rule of thumb, this "35 BTU/Hr./Sq. Ft.," and it's worked well over the years. If you allow a maximum of 35 BTU/Hr./Sq. Ft. you'll avoid "Sweaty Foot Syndrome" and uncomfortable clients.

Remember, if you stand on something that's hotter than you are, your body will retain too much of its natural heat and you'll feel uncomfortably warm. If the surface is the same temperature as you, you'll feel great.

Q: What determines the 35 BTU/Hr. per square foot?

A: It's a function of the spacing of the radiant-floor tubing, the temperature of the water and the material that makes up the floor.

Generally, radiant floors deliver about 2 BTU/Hr. per square foot for each degree F difference in temperature between the surface temperature of the floor and the air temperature in the room. So, for instance, if your floor is 85°F and your air temperature is 68°F (which feels more like 72° in a radiant job) the difference is 17°F (85 - 68 = 17). Figuring we have 2 BTU/Hr. per square foot, per degree F difference (17 in this case), we get 34 BTU/Hr. per square foot of radiant floor surface.

Q: The air can be that cool and I'll still feel comfortable?

A: You sure will! In fact, when you breathe that cooler air into your lungs you'll get the same feeling you experience on a crisp autumn day. The air feels fresher than it does in a building heated by any other means.

Q: But I won't feel cold?

A: No, because the surface temperatures in the room will control the heat loss from you body. You'll feel terrific.

Q: Okay, so once I have the radiant capability of the floor in BTU/Hr., what do I do next?

A: Compare this number to your amended (reduced by 15%) I=B=R or ASHRAE or ACCA heat loss calculation. If your "radiant heating capability" number is greater than your amended heat loss calculation number, you'll be able to meet the room's heating needs on the coldest day of the year solely with radiant-floor heat.

Q: Suppose my amended heat loss is *greater* than the radiant capability of the floor?

A: If the heat loss is greater than what the available floor space can provide (which often happens in rooms with three exposed sides and lots of glass), you'll have to figure on adding a supplemental source of heat. This could be panel-type radiators, a duct-coil, a toe-kick heater with a low-set aquastat, or any number of other hydronic heaters.

Q: How would I operate that supplemental source of heat?

A: The simplest way would be as a separate zone through a two-stage thermostat. The thermostat's first stage would run the radiant system; the second stage would kick in when the outdoor-air temperature dropped to design level, and then only if the radiant system couldn't keep up with the heat loss on that day.

Q: Could I add additional radiant tubing to the walls or ceiling instead of putting in a supplemental source of heat?

A: Yes, you could. Just watch out where the folks hang pictures or light fixtures. And keep in mind the owner will have limited options in the future if he should decide to move walls around.

Q: Is there a way to estimate how much radiant-floor tubing I'll need on a job?

A: A good rule of thumb is to figure on using 1.25 feet of radiant-floor tubing for each square foot of floor space (assuming the radiant-floor tubing is going in a slab).

This will allow you to place the radiant-floor tubing on 6" centers near the outside walls (for a distance of about 24" *in* from the wall), and on 12" centers for the rest of the room. This is typical residential spacing.

For instance, if you had a 20' x 20' room to heat, you'd need 500 linear feet of tubing (20' x 20' x 1.25 = 500').

Q: Why is the tubing closer together near the walls than it is in the center of the room?

A: Because most of the heat loss is near the walls.

Q: Do I start my first row of radiant-floor tubing right at the outside wall?

A: No! Always stay about six inches away from any exterior wall so the carpenters and rug installers don't drive nails through it.

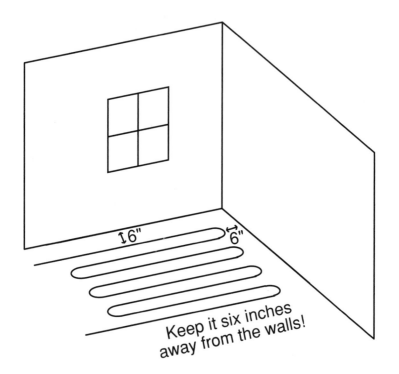

Q: Does the spacing you just described also apply if I'm stapling the radiant-floor tubing underneath a subfloor?

A: No, for these jobs, you'll sometimes have to place the radiant-floor tubing on 4" or 6" centers because the tubing is not surrounded by the flooring material. Keep in mind, as well, that the thicker the floor (and the carpeting), the closer together the radiant-floor tubing will have to be. The water circulating through the radiant-floor tubing will also have to be hotter if the floor is thick or carpeted.

Given that the size and heat losses of two rooms are equal, you'll **always** need more tubing on a staple-up job than you will on a slab job.

Q: How do I know how much radiant-floor tubing I'll need when the tubes are all on different centers?

A: Again, as a rule of thumb, you can use this chart to estimate the total linear footage of radiant-floor tubing you'll need when the tubes have to be closer or, in some cases, further apart:

When the Spacing is:	You'll use this many Linear Feet of Tubing per Sq. Ft. of Floor Space
4" on center	3.10
6" on center	2.05
8" on center	1.65
9" on center	1.38
12" on center	1.05
15" on center	0.83
18" on center	0.70

Q: When would I have the radiant-floor tubing on wide centers such as 15 and 18 inches?

A: The wider spacing I'm showing you on the chart would be for an area such as a garage, which you'd normally keep at a lower temperature than a living space.

Q: In residential work, how far apart would I usually place the radiant-floor tubes?

A: It varies with the room you're heating. As a rule of thumb, here are some common residential spacings:

Typical Residential On-Center Dimensions			
Room	**Thin Slab (2" slab on Grade**	**Slab below Grade**	**Staple-up**
Living	6" (2' in from ext. walls) and 12" (rest of room)	12"	8"
Dining	6" (2' in from ext. walls and 12" (rest of room)	12"	8"
Bedroom	12"	12" to 15"	8"
Bathroom	9"	9" to 12"	8"
Kitchen	12"	12"	8"
Laundry	15" to 18"	15" to 18"	8"
Study/Den	6" (2' in from ext. walls) and 12" (rest of room)	12" to 15"	8"
Int. closet	No Heat	No Heat	No Heat

Q: How do I get the water from the boiler to the radiant-floor tubing?

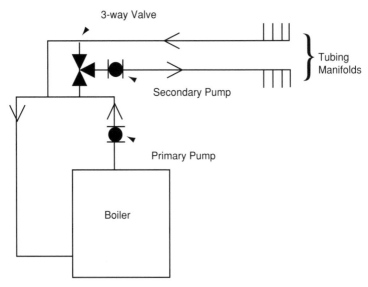

A: You supply it through either metal pipes or plastic or rubber tubing (of a larger size than you're using in the floor), and connect to a pair of manifolds.

Q: What do manifolds look like?

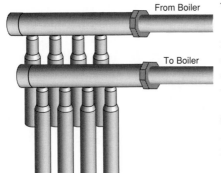

From Boiler

To Boiler

A: Like this.

You need a pair for each radiant heat circuit. One supplies the hot water, the other returns the cooler water to the boiler.

Q: How do the radiant-floor tubes connect to the manifold?

A: With either compression-type or barbed fittings. One end of the tube goes on the supply manifold, the other goes on the return. If you're installing rubber hose, don't use soap, oil or any petroleum- or silicone-based lubricant to help slide the hose onto the barb. Use rubbing alcohol instead. (Most of the installers I've watched simply spit into the end of the hose before sliding it onto the barbed fitting. So much for textbooks.)

Oh, and it's best to connect the tubes in a reverse-return pattern.

Q: What's reverse-return?

A: It's a method of piping that balances flow through all the circuits. In this case, the first tube supplied is always the last tube returned.

Q: Can you give me an example of this?

A: Sure! Here's how you'd pipe a manifold set with four tubes.

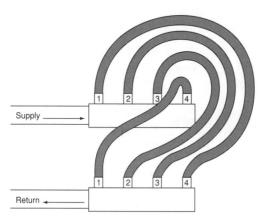

Notice how the supply side of tube #1 goes on manifold tapping #1 while the return side of that same tube goes on manifold tapping #4. Tube #2 connects its supply to manifold tapping #2 and its return to manifold tapping #3, and so on.

Set up this way, the distance to and from each tube through the manifold is the same, so you'll have an easier time balancing the flow.

Q: Do the tubes all have to be the same length?

A: Ideally, yes. If they're all the same length, the balance will be perfect. For instance, let's say your radiant heat zone needs a total of 600 feet of radiant-floor tubing. You could use a manifold set with three tappings and run 200 feet of radiant-floor tubing between each of the three tappings. That will give you a total of 600 feet.

If you wanted, you could use a four-tapping manifold set and run four, 150' lengths of radiant-floor tubing. The result would be the same.

Q: Suppose, for some reason, I have to use two, 250' lengths of radiant-floor tubing and one, 100' length. Will these balance?

A: Not without some help. In a case such as this, you should install a small balancing valve at the return side of the manifold tapping serving the shorter circuit. Then, all you have to do is add some resistance to flow through the shorter circuit to equal the pressure drop through the longer circuits.

Q: Isn't this similar to the way I'd balance a long and a short loop of baseboard radiation?

A: Yes, it's exactly the same.

Q: Do I need gauges to do this?

A: No, you can pretty much do it by touch. If you want to be more accurate, you can check the temperature from supply to return with a contact thermometer. Most people design radiant-floor systems to have a 20°F temperature drop from supply to return.

Q: Are the circulators for radiant-floor heat jobs different from the circulators I use on other types of systems?

A: No, they're the same. What you're looking for are small flow rates against relatively high pressure drops. The small, water-lubricated circulators you use from day to day will usually work well.

Q: How do I size one of these circulators for a radiant-floor heat job?

A: The same way you'd size it for any other hydronic system. Each zone usually gets its own circulator (although you can also use zone valves). You size the pump to the proper flow rate you'll need to deliver the BTU/Hr. to the zone (usually at a 20°F temperature difference). You'll also have to consider the friction loss through the highest-pressure-drop circuit.

Q: Is there a rule of thumb I can use for sizing circulators for residential radiant-floor heating work?

A: If you connect no more than five circuits to each manifold set (each no longer than 200 feet), you should be able to make the job work well with a small water-lubricated circulator that can move 5 gpm at about a 14' head.

Q: Does all of the return water go directly back into the boiler?

A: Not usually. It's best to bypass some of the return water back into the supply by using a two-, three-, or four-way valve.

Q: Why do I have to do that?

A: To make sure the return water entering the boiler isn't too cool.

Q: Why is that important?

A: Because cold return water can make the products of combustion reach their dew point and condense inside the boiler or the flue. This, over time, can erode the metal.

There's also the potential for thermal shock if cold water enters a hot boiler. Thermal shock is what happens when you hit hot metal with cold water. The sudden contraction of the metal can cause it to fracture.

The mixing valve also allows you to modulate the supply temperature heading out to the zone by blending cooler return water into that hotter boiler water. This way, you're able to supply different temperature water to different zones.

Q: How do I pipe a three-way valve into a radiant-floor heating system?

A: Like this:

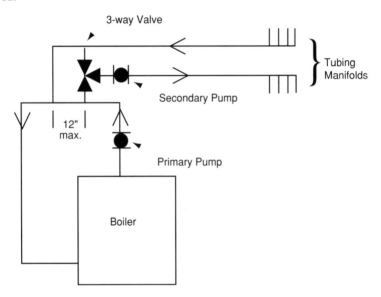

You set the valve to deliver the temperature you need to heat the zone. From there, the valve automatically mixes hot boiler water with cooler return water to give you the mix you're looking for.

Q: How about a four-way valve? How is that different?

A: A four-way valve looks like this in a system:

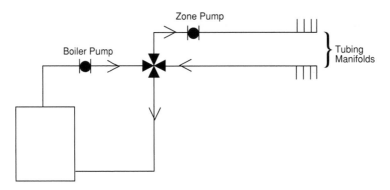

And here's what it looks like on the inside when it's in various positions:

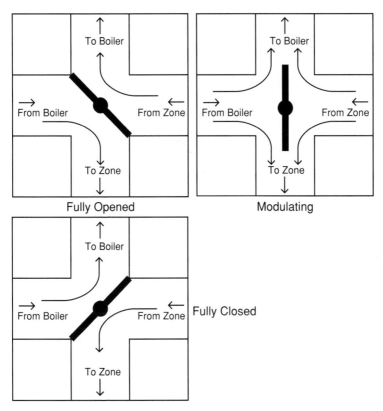

As you can see, there are four tappings on this valve. Two go to the boiler, two go to the radiant-floor heating zone. When the four-way valve throttles to a fully closed position, all the water flows through the circulator and back into the tubing. When the valve's in this position, it's as though the boiler isn't even there.

As the valve throttles toward its fully opened position, some of the hot boiler water flows out to the zone while the rest of the hot water flows back into the boiler to prevent condensation and thermal shock.

Q: Is the boiler pump necessary with a four-way valve?

A: It's most important when you're using a copper fin-tube boiler or a low-mass cast-iron boiler. Proper flow rates through these boilers are crucial.

Q: What about two-way valves? Can I use them as well?

A: Yes, you can. And here's how you'd pipe one in.

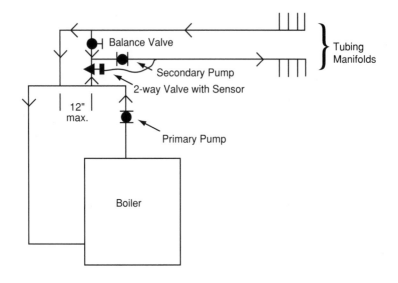

Q: How does this set-up work?

A: The two-way valve senses the water temperature flowing to the zone and closes on a rise in temperature. Take a close look at the diagram. The secondary, zone circulator comes on and moves hot water out of the primary, boiler loop. A portion of the return water moves through the bypass line and mixes in the tee to cool the supply water moving into the zone.

As the supply temperature rises, the two-way valve throttles, limiting the amount of hot water available to mix with the return. This sets a limit on the temperature flowing out to the zone.

Q: Is there an advantage to using a two-way valve over a three-

or four-way valve?

A: Two-way valves are usually smaller and less expensive. Most also have a locking and limiting feature that discourages tampering once you have the system set up.

Q: **Are all these valves always self-contained?**

A: No, you can add a motorized operator and a reset controller so the system becomes temperature responsive.

Q: **What's "temperature responsive?"**

A: This means the water temperature moving out to the zone responds to the outdoor-air temperature. When it gets colder outside, the water temperature going to the zone automatically increases. As it gets warmer outside, the supply water temperature gets cooler. We call this a "reset" system because the controls and valves "reset" the temperature in response to the weather conditions at any given moment.

Here's a typical heating curve a control manufacturer would use to match the boiler temperature to the outdoor-air temperature.

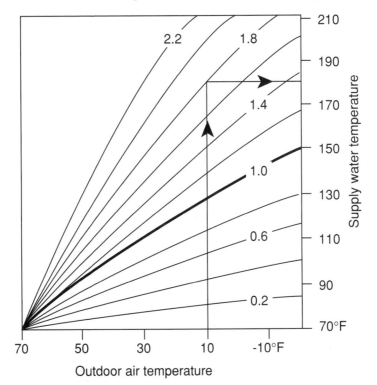

Q: **What's the advantage to this?**

A: You save fuel and gain closer control over the system.

Q: **What operates the motor?**

A: A small microprocessor that senses the water and air temperatures and then decides, based on a heating curve that you preselect, what the temperature of the water should be at any given moment.

Q: **Are these controls expensive?**

A: Somewhat, and for this reason, many installers put in residential radiant-floor heating systems without temperature-responsive controls.

Q: **How do they control those systems?**

A: With a room thermostat that operates the zone circulator, in the same way it would in a baseboard system.

Q: **How's the level of control on these jobs?**

A: It's not as fine-tuned as it would be with a temperature-responsive, reset system, but for most residential work, it seems to be fine.

Q: **What's the main difference between a job installed with a temperature-responsive control and one installed with, say, just a three-way valve and a circulator?**

A: In the temperature-responsive job, the circulator runs continuously while the water temperature varies. In the other, less-complicated, system, the water temperature is constant and the flow rate varies.

Q: **Is one way better than the other?**

A: If you can afford the controls, it's better to run constant circulation and vary the temperature of the water. A good rule of thumb is that the controls should not cost more than the tubing on any given job.

Q: **Are there areas of the country where reset works especially well?**

A: If you're working in an area where there are dramatic changes in the outdoor temperature from day to night (the desert and the mountains, for instance), you should use continuous circulation and reset control for the greatest level of comfort.

Q: Will I need any other special hydronic accessories for these jobs?

A: You'll need an automatic fill valve to set the initial system pressure. Remember to close the valve going to the automatic fill once you've set the system pressure. You'll also need a properly sized expansion tank. You may need a back-flow preventer, depending on the codes in your area.

I also think you should have a low-water cutoff on the system to protect the boiler against dry firing, regardless of whether or not the code calls for one.

You'll also need a strainer to keep debris out of the tubing. This is something you wouldn't normally use on a residential hydronic system, but here, it's a good idea.

Flow-control valves or spring-loaded check valves will keep heat from migrating into zones that are higher than the boiler. You may need two, one on the supply side and another on the return side, to prevent "gravity" circulation.

Q: Are there rules for zoning a radiant-floor heating job?

A: Yes. Rooms that are next to each other and have similar floor coverings and heat losses should be on the same zone.

An exception would be a master bathroom with a tiled floor and an adjacent master bedroom with a rug. These should be on the same zone because the tile naturally will feel warmer than the rug. In other words, the bathroom will feel warmer than the bedroom, which, in most cases, is exactly what you're looking for.

Each floor of the house should be on its own zone. Don't put the first and second floor, or the first floor and the basement, on the same zone.

Areas that have doors opening to the outside should be on their own zone because of the unusual heat loss. And for the opposite reason, kitchens and laundry rooms should ideally be on their own zone because of the heat gain from the stove, oven and clothes dryer.

In short, follow the same common-sense procedures you'd use when you're zoning *any* type of hydronic heating system.

Q: What about a four-season room?

A: These "sun rooms" present their own particular challenges and should always be on their own zone. The best way to sense the temperature in this type of space is with a slab-temperature sensor (you drill a hole in the floor for this one).

Don't use an air thermostat because on a sunny winter day the air in the room can get warm enough to shut off the flow of water to the slab. Once the floor gets cold, you'll fall into that "cold 70" trap where there's a great difference between the air temperature and the temperature of the surrounding surfaces. The result? Discomfort.

The same rules apply to any area that sees a lot of solar gain. These areas should always be on their own zone.

Q: When I first look at the job, are there things I should watch out for?

A: Yes, rooms with three sides exposed to the outdoors, particularly when they have a lot of glass, should always make you stop and think. These rooms often have heat losses through the floor and the ceiling as well. Chances are you'll need some supplemental heat to help out the radiant-floor tubing on a very cold day.

Q: How about rooms with a lot of stone? Can they present problems?

A: Yes, rooms with fireplaces, cooking hearths, brick walls and any other surface that can store heat for a long time can trick you. It takes more time for a room such as this to come up to a comfortable level, but once there, it will most likely stay comfortable longer after the burner shuts off.

These "stone rooms" also should be on their own zones.

Q: Do fireplaces present any other problems?

A: A fireplace is romantic, but it can suck heat out of a house like a giant vacuum cleaner (the same goes for exhaust fans in kitchens and bathrooms).

Watch out for fireplaces and take them into consideration when you're setting up the zones. They won't necessarily affect the comfort level in the room they're in, but they *will* draw heat away from adjacent rooms and can make them feel uncomfortably cool.

Q: Suppose the house is near the ocean. Can that be a problem?

A: Not if you plan for it. If the house sees prevailing winds over 7 mph, it will have greater-than-normal, wind-induced infiltration losses. That 15% you took out of your I=B=R heat loss calculation may have to go right back in when you're near the ocean. Use your common sense here.

The same rule applies to a house on the top of a hill.

Q: Is there anything else I should watch out for?

A: Yes, recessed lighting. By code, "high hats" can't be insulated (although several companies now offer pre-insulated "high-hats" that do meet the code). When they're not insulated, these lights act as little "chimneys," pulling BTUs out of the room into an unheated attic space. Since they increase infiltration, they can have an effect on your sizing, so watch out for them.

The same thing goes for the sheetrock around support posts. Feel around them with your hand. They, too, may be sucking heat up into an unheated attic space.

Houses that are over 50 years old are not at all well insulated. Your heat loss will be greater in these "old barns," and you may need supplemental heat.

If you're putting radiant heat in the floor of a bathroom, keep the tubing at least a foot away from the toilet's wax gasket.

Look around and use your common sense.

Q: If I'm stapling the tubing to the bottom of the floor, do I have to watch out for anything in particular?

A: You should always make sure the floor boards are as dry as possible. Ideally, you should allow the wood to dry inside the house while you run the heating system under the subfloor for several days.

Wood floors that contain more than eight to ten percent moisture can shrink or bow when the system first comes on if you don't take proper precautions.

Make sure your customer (and the builder) knows what's at stake here. Use a moisture detector to check the wood.

Q: Is it important for me to know what materials they used to build the floor?

A: It's crucial. Some floors have several layers of wood. The thicker the floor, the tougher it is to push BTUs through it. Some subfloor materials don't transfer heat well at all.

Know what you're dealing with and consult with the tubing manufacturer on the best course of action for that particular job.

Q: Are there special wooden floors that are more compatible with radiant-floor heating systems?

A: Yes, they're called "floating" hardwood floors. They fit together in a tongue-in-groove fashion, but they're not attached to the subfloor in any way. The base molding covers a small expansion gap between the edge of the floating floor and the walls. The floor is free to expand and contract without a problem.

Q: How do I run the tubing when I'm attaching it to the underside of a floor?

A: The simplest way is to drill a 1¼" hole through the center of the joists at one end of the basement. Make sure you drill into the **center** of the joist, not at the top or bottom, and make sure the holes line up. Then work off your zoned manifold sets, threading the tubing back and forth between the joists.

Here's what the under-side of the floor will look like as you install the tubing.

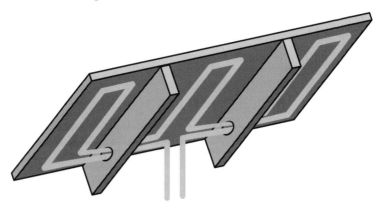

Q: Do I have to insulate under these tubes?

A: Absolutely. But leave at least a two-inch air gap between the tubing and the foil face of the insulation so the radiant waves of energy can bounce off the foil and up into the floor. This will help spread the heat more evenly across the floor and prevent striping.

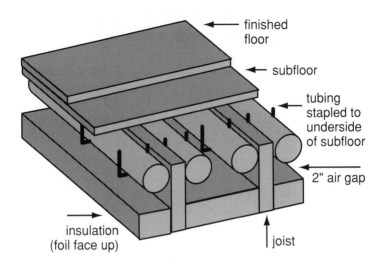

finished floor

subfloor

tubing stapled to underside of subfloor

2" air gap

insulation (foil face up)

joist

Q: Wouldn't it be better if I just pressed the insulation up against the tubing?

A: No, it would be worse because the rays of radiant energy wouldn't be able to diffuse evenly under the floor. You'd be able to feel where the tubes are when you walk on the floor.

Q: How do I attach the tubes to the underside of the floor?

A: You can simply staple them directly to the floor. If you're using plastic tubing with an external oxygen-diffusion barrier, use the special staples the manufacturer provides so the barrier doesn't get damaged as the tubing expands and contracts.

Place the staples every six inches so the tubing doesn't sag and break contact with the floor. Make sure the staples don't crush the tubing and impede the flow of water. Most of the tubing manufacturers can rent you a staple gun with a guard that sets the staples at the right depth. You can also use the heat diffusion plates we looked at earlier to attach the tubing.

Q: Is radiant-floor heating a good choice for a vacation home in a cold climate?

A: I don't think it's your best choice if you want to arrive on a frigid Friday night and have the place warm and toasty in fifteen minutes. Radiant-floor systems take a while to get started, but then again, they keep the place warm *long* after they're turned off.

For a vacation house, you should have a supplemental convective heat source to take the chill off while the radiant-floor system comes up to speed. A duct coil or a kick-space heater working off a two-stage thermostat does a good job.

Q: How does a radiant-floor heating system affect the humidity level in a house?

A: It has relatively little effect. The lower air temperatures usually mean the relative humidity will be slightly higher than they would be in a convective hydronic system.

Q: Can I air-condition with a radiant floor or ceiling by circulating cold water instead of hot?

A: It depends on where you are in the country. If the relative humidity is high, the cold floor or wall surfaces will become damp (and dangerous if it's a floor). Some contractors in the western states where the relative humidity is low have been experimenting with radiant cooling with greater success.

Q: Did they ever use steam for radiant heat?

A: They tried in the early days, but with no success. Steam is simply too uncontrollable for this application.

Q: Did the old-timers ever try to install radiant systems without circulators?

A: No, the runs are too long for gravity circulation. To make a system work without a circulator, the pipe size would have to be so large it would become very unpractical.

Q: Is this the ultimate heating system?

A: Yep!

CHAPTER SIX:

CONDENSATE
HOT-WATER
HEATING

"Dan, I'm going to let you in on a few tricks I discovered along the way, but you have to promise me you'll pass them on."
—Mr. John Rogers, P.E.
(You would have liked him a lot.)

Q: Is it possible for me to use the condensate in a steam boiler to make a hot water zone?

A: Sure! Heating professionals have been doing this for years in areas such as New York City where there's still a bodacious amount of steam heat.

Q: Do I need a water-to-water heat exchanger to do this?

A: You *could* use a water-to-water heat exchanger or the boiler's tankless coil if you'd like. Either will give you a first-class installation. However, if you follow a few simple piping rules, you can get the job done *without* the heat exchanger.

Q: Let's take this one step at a time. First, how would I do the job *with* a heat exchanger?

A: You'd set it up much the same way you would any hot water system. The water-to-water heat exchanger would become the "boiler" for your hot water zone.

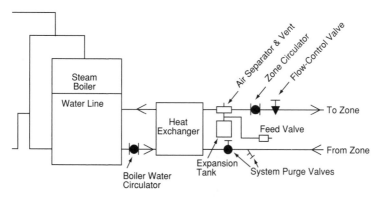

Pressurize the new zone with an automatic feed valve and a properly sized compression tank. Wire the circulator to come on at the call of the room thermostat. Have it pump away from the compression tank and toward the radiation.

You'll also need a second circulator to move water from the boiler to the heat exchanger. Wire this one so it runs at the same time as the zone circulator. Use a switching relay and a high-limit aquastat to control the burner during times when you're not making steam.

Q: Is it absolutely necessary for me to have that second circulator between the boiler and the heat exchanger?

A: The system works best if you do, but you could also let hot boiler water circulate through the heat exchanger purely by gravity. If you do it this way, however, the system's response to a call for heat won't be as fast as it would be if you use the second circulator. That second circulator pays for itself in system performance in no time at all.

Q: Are there any other drawbacks to circulating the boiler water through the heat exchanger strictly by gravity?

A: The main drawback is you need full-size tappings in the boiler. The size you need is typically 2" and the trouble is most modern steam boilers don't give you many extra tappings, let alone 2" ones.

Q: What happens if I use smaller boiler tappings?

A: The resistance to flow through the heat exchanger will be greater, so less hot boiler water will flow through the heat exchanger. Less flow means less heat transfer. You may not be able to get enough heat up to that new zone.

Q: Who manufactures a good heat exchanger for this purpose?

A: Everhot (191 Arlington Street, Watertown MA 02172) makes one that I've seen work well on many jobs. Their unit is very similar to the old tank-type heaters from the Steam Era.

The Everhot Model RH-8 will handle a hot water radiation load of about 45,000 BTU, which is more than you can expect to deliver through a ¾" line to a hot water zone.

Q: Does the heated water going out to the zone flow through the coil or through the tank?

A: The zone water goes through the coil. The boiler water flows through the tank.

Q: Does the hot water radiation have to be lower than the steam boiler's water line?

A: No. If you're using a heat

exchanger, your compression tank will keep the system under pressure. The fill pressure and tank size determine how high you can place the radiation above the boiler's water line. The only limit is the working pressure of the equipment you're using. It takes 1 psig water pressure to raise water 2.31 feet straight up. So, if you have equipment rated for, say, 100 psig, you'll be able to lift water about 230 feet to a hot water zone—if you ever wanted to (I don't think you'd ever want to).

Let me put it another way—that two- or three-story house won't be a problem.

Q: Suppose I decide I don't want to use a heat exchanger, will my radiation have to be lower than the steam boiler's water line?

A: No. If you use ¾" supply and return piping, **and make sure you don't use any air vents whatsoever in the zone piping**, the radiation can be as high as 30 feet above the steam boiler's water line.

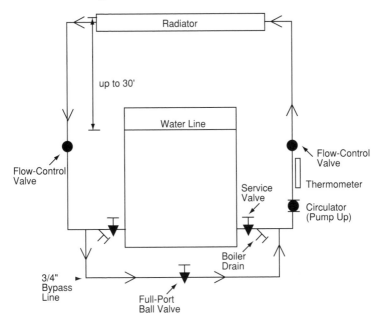

Q: What keeps the water up there in the zone piping if there's no automatic fill valve or compression tank?

A: Atmospheric pressure. It's the same phenomenon that keeps water in a straw when you put your finger over the top end and lift it out of a glass of water.

The water tries to fall out of the straw, but the atmospheric pressure (the weight of the air) pushes it back up. Since no air pressure can get into the top of the straw to balance the air pressure at the bottom of the straw, the water just hangs there.

Q: But a ¾" pipe is wider than a straw. Won't the water fall out of it?

A: Not unless air gets up into the zone. The principle is the same, regardless of the width of the pipe. At sea level, the atmosphere pushes down on everything with a pressure of 14.7 pounds per square inch. One pound per square inch can lift water 2.31 feet straight up. So, if you have a pipe that's sealed at the top and completely filled with water, the atmospheric pressure will be able to support a column of water about 34 feet high (14.7 psi x 2.31 feet = 33.957).

It has to do with pressure, not the width of the pipe. Get it?

Q: I'm not sure. Can you give me another example?

A: Here, think about this. Suppose you were to lift an upended glass out of a bucket of water.

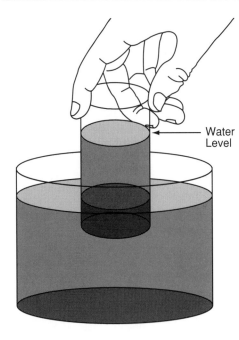

Water
Level

The water stays in the glass because the atmospheric pressure pushes it up there. The glass is certainly wider than a straw, but since air can't get into the top of the glass to offset the pressure at the bottom, the water stays where it is.

For practical purposes in your hot water zone, you can raise water 30 feet (or about three stories) and make it stay up there.

Q: **Suppose I live in Denver, Colorado? Will this still work?**

A: Yes, you just won't be able to keep the water up as high because the atmospheric pressure is less in Denver than it is in, say, New York City. The atmospheric pressure in Denver is about 12 psi. One psi still lifts water 2.31 feet, so let's see…12 psi x 2.31' = 27.72.

In Denver, your zone could be, say, 25 feet above the boiler water line. That's enough to get you up to the second floor of a house with no problem at all. Going to the third would be cutting it close.

Q: **But I can't use any air vents at the top of the zone, no matter where I live, right?**

A: Right, an air vent will let air into the top of the system. Once that happens, the water will fall back into the steam boiler.

Q: **Does this apply to manual air vents as well as automatic air vents?**

A: Yes, it does. You want to have the tightest possible connections you can get above that steam boiler's water line. For this reason, you should also avoid using any valves with packing glands. They, too, can draw air into the system and cause the water to fall out of the zone.

Just solder the joints tightly and let it go at that.

Q: So how can I fill the zone with water if I can't vent the top?

A: Set it up as a loop system and fill it with a garden hose.

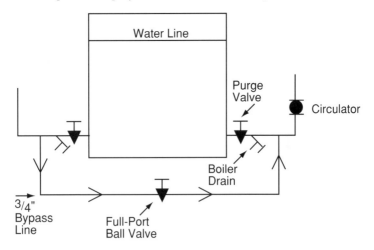

Pipe in two ball valves (or gate valves) and two boiler drains below the boiler water line. Shut both ball valves (or gate valves). Now connect the garden hose to one of the boiler drains and let the other flow freely to a drain.

Fill the zone piping under city water pressure until all the air comes out. Then, shut off the discharge boiler drain and the inlet boiler drain (in that order). Now open the two ball valves (or gate valves). If you did a good job of soldering, the atmospheric pressure will keep the water up in the zone.

The atmospheric pressure keeps the water from falling out!

Q: Are you sure of that?

A: Yes! Let's go over that basic principle one more time.

Get yourself a glass and a stiff piece of cardboard. Fill the glass to the brim with water and place the cardboard on top of it. Hold the cardboard and turn the glass upside-down.

Now, let go of the cardboard.

It stays there, right? That's because the atmospheric pressure is stronger than the column of water in the glass. This is the same principle that will keep the zone filled with water when you disconnect the hose. Trust me, it works.

Q: If I'm not using a heat exchanger with a compression tank will there be any pressure on the water that's in the zone?

A: There won't be any pressure at the *top* of the system. There will, however, be pressure on the water further down in the zone.

Q: Where does this pressure come from?

A: This is the static pressure we talked about before. It comes from the weight of the water. The higher the column of water, the greater the static pressure at the bottom will be.

Sea level

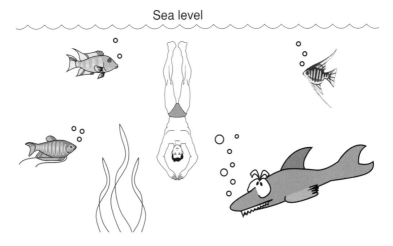

Want another example of static pressure? Think of the water in the ocean. There's no pressure at sea level, but as you dive deeper and deeper, the pressure increases because there's more water sitting on top of you.

In our zone, "sea level" is the high point.

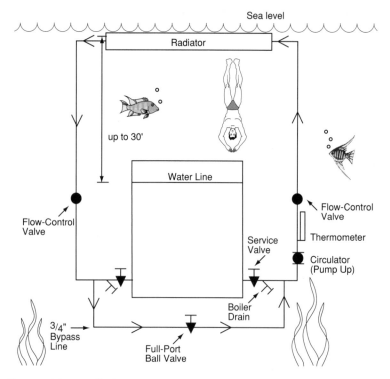

Q: How hot will the water in my steam boiler usually be?

A: It depends on the steam pressure. The boiling point of water increases as the pressure on top of the water increases. For instance, if your steam boiler operates at 2 psig pressure, the water in the boiler will boil at 219°F. If you run the boiler pressure up to 5 psig, the water won't turn to steam until the temperature reaches 227°F. At atmospheric pressure (which is what you have at the top of the system), water boils at 212°F.

Q: How does this affect the operation of my hot water zone?

A: If you're not using a heat exchanger and a compression tank, the water at the top of your zone can boil if it gets too hot.

Q: Will this be a problem?

A: It sure will! When water boils, it "flashes" into steam and increases in volume about 1,700 times. It does this instantly and, in a sealed "container" such as a piping system, it does it with great violence. "Flash" steam is potentially very dangerous. It could actually blow soldered joints apart.

Q: When is this likely to happen in my hot water zone?

A: When the zone circulator shuts off.

Q: How come?

A: The water usually can't flash to steam when the circulator is on because the circulator adds a certain amount of pressure to the water. That extra pressure can keep the water in the liquid state. However, when the circulator shuts off, its pressure disappears and that's when the water at the top of the zone can flash into steam.

Q: Can this happen if the zone is lower than the boiler water line?

A: It's possible, but it's not likely because the higher water level in the boiler puts a certain amount of static pressure on the zone.

Q: Well then, how can I make sure the water at the top of my system never flashes into steam?

A: By making sure the water entering the zone never approaches 212°F. The simplest way to do this is to blend a portion of the water that's already been through the zone with the hot supply water leaving the boiler.

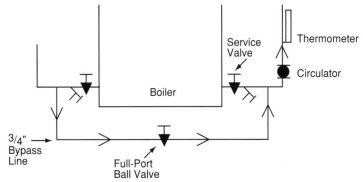

Q: Do I need special valves to do the mixing?

A: Not at all. All you need is a ¾" copper line between the zone return and the inlet side of the circulator, a ¾", full-port ball valve and a thermometer.

Pipe the ball valve into the bypass line and leave it fully open when you first fire the steam boiler. Next, let the boiler come up to steam pressure, and then start your circulator. With the bypass ball valve fully open, nearly all the water in the zone will bypass the boiler because it's easier to go through the bypass than it is to go through the boiler. That's the path of least resistance.

Now, to make the water flowing into the zone hotter, all you have to do is throttle the bypass ball valve a bit. Keep your eye on the thermometer as you do this. You'll see the temperature rise very quickly.

Q: Why does the temperature go up?

A: When you throttle the bypass ball valve, some of the returning water finds it easier to flow through the boiler than it does to go around the bypass.

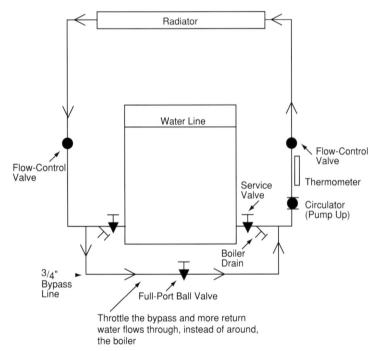

When this hot water comes out on the other side, it mixes with the water that bypassed the boiler, and gives you a mix that's hotter than the return water, but cooler than the boiler water.

Q: How much water should I let through the boiler?

A: The flow in gallons per minute isn't crucial here. Just keep your eye on the thermometer on the line supplying the zone. Stop blending when it reaches 180°F. Then, take the handle off the ball valve so no one else can mess with it. This will be the hottest water the zone will ever see.

Q: So as long as the supply water going up to my hot water zone never reaches the boiling point, it can never flash into steam?

A: That's right. You've set a fixed limit on the temperature by mixing the return water into the supply water while the boiler was making steam at its pre-set pressure.

Q: Is there anything that can change that temperature?

A: Well, if someone raised the steam boiler's pressure the water temperature in the boiler would also go up. But by setting the blended water temperature at 180°F, you've left yourself a comfortable margin of safety.

Q: Does it matter whether the circulator pumps away from the boiler or toward the boiler?

A: It's better to pump away from the boiler.

Q: How come?

A: Since this isn't a pressurized system, there's always the possibility the water can flash to vapor inside the circulator if the pressure drops too low when the circulator comes on. Technically, this is called "cavitation" and it can destroy a circulator in no time at all.

Q: Why does the pressure drop when the circulator comes on?

A: Because the circulator throws out what's inside of itself. This causes an immediate drop in pressure at its suction. This centrifugal action is what makes the water flow into the circulator in the first place. The trouble starts when there's not enough pressure behind the water flowing into the circulator to keep it in a liquid state.

Q: So the circulator itself can change the boiling point of the system water?

A: In effect, yes. It can change the boiling point of the water that's entering the circulator. And keep in mind, you don't have much pressure to work with on these jobs—just the height of the water in the boiler above the circulator. And then there's the pressure drop of the piping between the boiler and the circulator to consider.

If you start off with a limited static pressure (the height of the water in the boiler) and then lose a portion of it to friction in the approach piping, it may be difficult to control things when the circulator starts.

That's why it's best to pump away from the boiler. By keeping the circulator close to the source of the pressure (the water in the boiler), you lessen your chances of having problems.

Q: Does this apply to one case more than another?

A: It's especially important for those zones where you have all the radiation and piping below the boiler water line. If you have the circulator on the return, the pressure drop through the piping can cause the boiling point to drop. Combine this with a high starting temperature (which you'd have if the steam pressure were high and if you decided not to use that bypass blending valve), and you're asking for trouble.

In the case where you have the zone radiation and piping *above* the water line, the circulator location isn't as critical because you have the static pressure of the water in the zone working in your favor.

It's good to get into the habit of putting the circulator on the discharge side of the boiler, though. That way, you'll always be in good shape, whether you're piping a condensate zone or *any* hot-water zone. Circulators always work best when they're "pumping away."

Q: Where should I pick up my supply tapping?

A: That will vary from boiler to boiler. Modern steam boilers don't have very many extra tappings. You always want to pick up tappings that are lower than the boiler's water line, of course. And remember to keep your circulator as low as possible to take advantage of the static weight of the water in the boiler.

If you have a blank tankless coil plate, you could drill and tap it at a low point for a ¾" supply. That works well.

Q: Could I use the steam boiler's lower gauge glass tapping?

A: No! If you pump water through this connection you'll never know where the water line in the boiler is. You'll also affect the low-water cutoff if it's hooked into the gauge glass on quick hook-up fittings. In addition, you can suck air into the zone piping.

Q: How about on the return side of the zone piping? Where does that tapping go?

A: Again, use any available boiler tapping below the water line (***never*** return water above the water line). If you can't find a return tapping, return into the system's wet return, near the boiler.

Q: Does it matter which side of the Hartford Loop I tap into?

A: It's best to tap into the boiler side and keep that return tapping well below the boiler's water line.

Q: Can I supply and return from the same side of the boiler?

A: You shouldn't enter and leave through the same section because the zone water won't have enough time in the boiler to pick up the necessary heat. Ideally, you should pipe from opposite sides in a diagonal pattern. Like this.

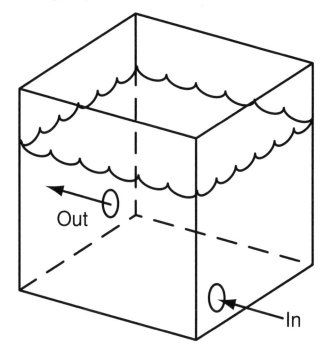

Q: Can I pipe straight through the boiler's mud leg?

A: No, because the water will zip through the boiler too quickly. It won't pick up enough heat, and you'll have a call-back for sure.

Q: Suppose there's no way I can tap directly into the boiler. Does that mean I can't zone with condensate?

A: You can still do it, but you have to use a bit of creativity. Here, just follow this diagram.

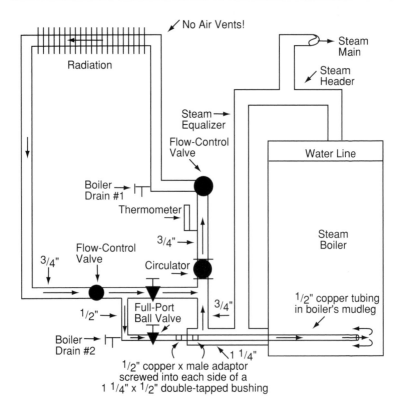

1/2" copper x male adaptor
screwed into each side of a
1 1/4" x 1/2" double-tapped bushing

Since you can't tap directly into the boiler, you'll use the 1¼" tee at the bottom of the steam equalizer line as both a supply and return point. Remove the plug from the existing 1¼" steam tee and replace it with a 1¼" nipple. Take a 1¼" x ¾" tee and screw a 1¼" x ½", double-tap bushing into it. Next, solder a length of ½" copper tubing to a copper x male adaptor and screw that into the *boiler* side of the bushing. Now, screw another copper x male adaptor to the other side of the bushing. That's your return tapping. Attach the tee with the tubing and bushing to the new 1¼" nipple.

The copper tubing reaches deep into the boiler's mud leg and deposits the cooler return water on the side opposite the one from which you'll be drawing your hot supply water. You'll take the hot supply from the bull of the tee. By doing it this way, you'll get the circulation you need across the boiler to pick up the heat for the zone.

Q: Do I still need a bypass line if I pipe it this way?

A: Yes, the bypass allows you to temper the water as it leaves the boiler. The bypass keeps the water from flashing to steam when the circulator shuts off.

Q: How do I start this one up?

A: Fill the zone with water by using the two boiler drains. Put a hose on #1 and purge back through #2. Let the boiler steam, and then start the circulator. Use your two ball valves to blend the water through the bypass until your supply temperature reaches 180°F. Take the handles off the ball valves and you're all set.

Q: Before, you said I shouldn't pipe directly through the boiler's mudleg because I wouldn't pick up enough heat on each pass. Does that happen with this system?

A: No, it doesn't because the copper tubing injects the water deeply into the boiler and forces it to reverse direction before it can leave the boiler. This action makes the return water at the bottom of the boiler very turbulent. It mixes things up, allowing you to pick up the heat you need for that zone.

Q: With any of these systems, should I put a strainer on the inlet side of the circulator so I don't suck mud out of the boiler and into the circulator?

A: No, because the strainer can produce a very high pressure drop at the circulator's inlet, especially when it gets dirty. That drop in pressure can make the circulator cavitate.

Q: Well, what's to keep the circulator from clogging up then?

A: If you choose the right circulator for this application, you won't have a problem. I like to use a 1,750 rpm, three-piece circulator for these zones because they have larger and wider impellers than their smaller, high-speed, wet-rotor cousins. A larger circulator fares better on this application because it can pass more debris through its impeller.

Q: Should I use an iron-body circulator?

A: You could, but a bronze circulator will last much longer here. Condensate usually has a good deal of carbonic acid mixed in with it. Bronze is a much better material than iron for this service.

Q: How should I control the zone?

A: The simplest way is to use a room thermostat to operate the circulator through a double-pole, single-throw relay such as Honeywell's R845A. You'll also need a single-pole, single-throw, immersion-type aquastat (such as Honeywell's L4006A) to set the maximum boiler temperature when you're not making steam.

Wire the controls so the circulator and the burner come on at the same time. The burner will bring the boiler water up to 180°F and no higher. The aquastat sees to that. It will shut the burner off, but the relay will keep the circulator running as long as the room thermostat continues to call.

By setting it up this way, you're able to supply hot water to the zone without making steam. As far as the folks upstairs are concerned, the steam system and the hot water zone are totally independent.

Q: What if the steam system recently shut off and suddenly the hot-water zone comes on. What happens then?

A: The boiler water temperature will be higher than the 180°F aquastat setting, so the zone thermostat (working through the relay) will start the circulator, *but not fire the burner*. Simple, isn't it?

Q: Could I pipe an indirect domestic hot water heater off a steam boiler using the same piping techniques I used for the heating zone?

A: Yes. Just treat the indirect heater as you would a radiator. Make sure you blend return water into the hot boiler water to limit the supply to the heater to 180°F.

Here's a sketch.

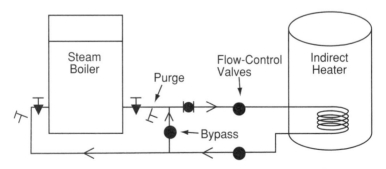

Q: How many hot water zones or indirect heaters can I take off a steam boiler?

A: That depends on the boiler's capacity. You can't take out more BTUs than you put in. I've seen people add one too many hot water zones and when it came time to make steam, they were out of luck.

Q: What's a practical limit for a house?

A: Again, a lot depends on the size of the boiler. Usually, you can get away with a ¾" line, flowing about 4 gpm. That'll deliver 40,000 BTUh, or so, to the zone. That's enough to heat a good size zone.

Q: Will I always be able to get 40,000 BTU/Hr. out of a residential boiler?

A: Yes, if the boiler's D.O.E. Heating Capacity load is 120,000 BTU/Hr. or higher.

Q: How come?

A: Because you're playing with the boiler's pick-up factor. The pick-up load usually represents about a third of the net load, depending, of course, on how the installer sized the steam boiler. One-third of 120,000 BTU/Hr. is 40,000 BTU/Hr. That's equal to 4 gpm, about what you can expect to flow through a ¾" copper line.

Q: What is the purpose of the pick-up load?

A: The pick-up load gives you the "extra" capacity the boiler needs to heat the pipes as the steam heads out toward the radiators.

Q: Is the pick-up load always available to me?

A: No, it becomes available to your hot water zone *only* after the steam pipes have heated up.

Q: How about if I'm not making steam?

A: If you're not making steam, the pick-up load (and the rest of the boiler load) is available to your hot water zone.

Q: So the pick-up load sets the limit on what I can do with this zone?

A: Yes.

Q: Let's say I'm sizing a new steam boiler and I want to use one or two hot water zones. Should I add the BTU/Hr. of the hot water zones into the BTU/Hr. load I need for the steam system, and then select a boiler for the combined total?

A: No! Never add the hot-water-zone load to the steam load when you're sizing a replacement steam boiler. This is a game of *subtraction*, not addition. Size the steam boiler first. Base it on the connected steam radiation load, plus a suitable pick-up factor, not the heat loss of the building. Then work with the available pick-up load to size your hot water zone or zones. You can't take out more than what's there.

Q: What happens if I oversize my steam boiler?

A: You'll have problems with the steam side of the system: surging water lines, water hammer, uneven heat, high fuel bills.

Q: Could I wire the boiler for priority? You know, either make steam or run the hot water zones, but not both at the same time?

A: Yes, but this limits the usefulness of the hot water zones. Suppose you want some heat but you can't get it because the steam zone is calling? Or vice versa.

Q: Can I set up a radiant-floor heating zone using the condensate from the steam boiler directly into the tubing?

A: I wouldn't. The water in the steam boiler is usually pretty dirty. Since the runs in a radiant-floor heating system can be 200' long (or longer) there's a good chance they'll get clogged. Use a heat exchanger instead.

Q: You mentioned Mr. John Rogers P.E. at the beginning of this chapter. Who was he?

A: John was the man who taught me these and many other tricks of the trade. I had the pleasure of working with him for a bunch of years once upon a time.

John had little formal schooling but nevertheless managed to educate himself to the point where he left home one day and took the New Jersey State licensing exam for Professional Engineers. He passed on his first try. Just blew that test away.

John did this just before he died on December 16, 1987. It was an exclamation point on a brilliant career. You would have liked John a *lot*. One day I was asking him a question about heat transfer, and before I knew what was happening, he launched into a discussion of a hummingbird's heart and how effectively that tiny pump managed to move the heat away from the little bird's body. Veins and arteries became water pipes, muscles became little boilers, blood became water.

John made me see the magic. He was the best overall heating man I've ever known. A *very* cool guy.

AFTERWORD

IF YOU HAVE IT IN YOUR HEAD...
YOU OWN IT

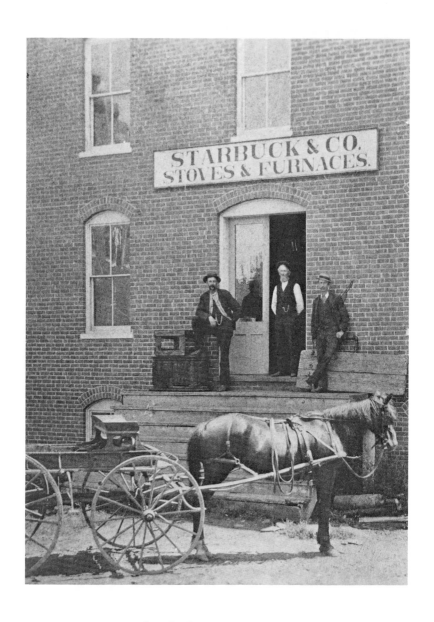

George Starbuck Jr.

George Starbuck Sr.

R.M. Starbuck

THE CARETAKER

*This story first appeared in the January, 1994
issue of **Plumbing & Mechanical** magazine.
My telling of it does not begin to do justice
to the family that lived it.*

He slid up to the registration desk like a truck coming in for
gas. "Doug Starbuck," he said offering me a hand that looked like
a calloused catcher's mitt. I shook it and checked his name off
the list for the steam seminar I was about to conduct.

"Starbuck," I said. "That's a familiar name."

"Maybe you know R.M. Starbuck's books?" he asked.

"I sure do!" I said, suddenly making the connection. "I have a
bunch of Starbuck books in my collection. He pioneered those
wonderful question-and-answer books for the trade."

"That's him," he said. "R.M. Starbuck was my great-granduncle.
He lived and worked right here in Hartford, Connecticut. His
brother was George Starbuck, my great-grandfather. And *their*
father was George Starbuck, Sr., my great-great-grandfather.
George, Sr. started our family business up in Turners Falls,
Massachusetts back in 1872. I'm still working that business. You
ought to come see it sometime; I'm still in the same building. Got
all this old stuff laying around, books and tools and stock. *Real*
old stuff."

The registration line was backing up, and one of the other guys
had to jump in to help. I didn't notice. I was floating back a hundred years or so with Doug Starbuck, The Caretaker of Heating
History.

You ride Interstate 91 straight up through Massachusetts, almost
to the New Hampshire line. Get off the main highway and drive a
few miles through the woods until you reach the Connecticut
River. Cross the old bridge down near the mills and head for the
center of town. You'll see Starbuck's shop on the left.

Doug was standing out in front, a bear of a man, waving his
beefy arm and smiling at us from beneath the George Starbuck &
Sons sign.

"Is that him?" Bob Steinhardt asked.

"That's him," I said, turning the van around. I'd brought Bob
along because we'd grown up together in the business, and he

was the first guy to show me a Starbuck book. Bob and I have
been in a lot of basements together and, from the way Doug
described it, I thought Bob would get a kick out of this place.

"Right on time!" Doug said as I made introductions. "C'mon in
the back."

We walked down the alley to a museum-piece of a building.
"This is the first shop," he said, waving his arms to take in the
place. "Just this part in the back here. We built this part in
1872." He pointed up the alley toward the street. "The front part
we built in 1888 after the business had grown. I rent part of that
space to the barber now. I'm the whole business now. I got all the
space I need back here."

A weathered wooden shed connected the first floors of the two
old brick buildings. Bob and I looked at each other as Doug wres-
tled with an old lock. "We built this shed to connect the buildings
years ago," Doug said, and then backing up and forgetting the
lock for a moment, he pointed up toward the second floor. "See
that pulley up there?" We nodded. "We used that for the stoves."
He used the word "we" as though the Dead Men were waiting
inside for us.

"What stoves?" I asked.

He chuckled. "That was the business back then, Dan. People
heated with coal and wood stoves. We'd go to their houses in the
spring, pick up their stoves, bring 'em back here, hoist 'em up to
the second floor, clean 'em, retube 'em, cover 'em with newspa-
per. Then in the fall, we'd take 'em back with the horse and
buggy and reinstall 'em. It was a good business."

Bob and I looked at each other. Doug stuck his key back in the
door lock and gave it another twist. "Here we go!" he said, pushing the old door inward. "Welcome to Starbucks!"

We followed him through. Right into the past.

It's the smells you notice first. It's like an old bookstore smell,
but there's something else there too. There's an oil and old-metal
smell mixed in as well. And sweat from 121 years of work. And
it's cold in there because of the tons of iron and steel and brass
piled everywhere. And it's quiet.

"My pipe machine can cut and thread up to eight-inch," Doug
said, pointing to this oily rhinoceros-like behemoth in the corner.
"Most guys don't do that anymore, but *I* can if I want to." He let
loose a child-like laugh that was contagious. "Most guys have
never even *seen* one of these!" He patted the machine. Bob and I
looked at each other and followed him into his office.

Doug has ancient column radiators fed with Honeywell
"Unique" valves. These valves stopped my legs from moving.

They stopped Bob's too. "What's wrong?" Doug asked with a sly smile.

"I've seen those things in books, books from around 1900, but I've never seen them installed."

"Those?" Doug said with a casual shrug. "I've got *plenty* of those in stock. How many you want?" He laughed again. I just shook my head in amazement because—and this is where you have to appreciate history—the "Unique" valve was one of Honeywell's very first inventions. They built an empire around these things.

The valve has an inlet, an outlet and one connection to the radiator. Dead Men used to use these valves to control their gravity hot-water systems. They did this a hundred years ago, and Doug Starbuck was still doing it today.

"You have these in stock?" I asked.

"Sure do! Got a basement full of them. How many you need?" He smiled again. "I'll take you guys down there in a minute, but first, I thought you might want to have a look at these."

He got up from his roll-top desk and walked over to a corner where he had a big cardboard carton filled with books. "Take a look," he said. "I put these together for you last night. Thought you'd like to see them."

The box was filled to the brim with old textbooks and catalogs. Bob and I looked at each other. "We don't like to throw stuff out," Doug said. "We save everything." He spoke for five generations.

"I see that," I said, as I lovingly flipped through the pages of a first-edition Hoffman Handbook. "You have more of this stuff?"

"Oh, the place is filled with all sorts of books," he said. "But you haven't seen the best part yet. C'mon."

He walked us down the hall to a staircase. "Go on down, I'll be right behind you," he said. Bob and I climbed down into a dimly lit stock room. There were rows upon rows of bins filled with fittings and valves and plumbing-and-heating specialties from a forgotten time. Each bin had a hand-lettered card that had to have been placed there 100 years ago. Bob and I looked at each other.

"How about this?" Doug grunted as he picked up one of the fittings from off the floor.

"What the heck is that?" I asked.

"It's a cross. You never seen a cross before?"

"I've seen a cross, but not like that!"

"What's so strange about this?" Doug asked, a gleam in his eye. He pointed to each opening as he spoke. "It's a six- by one-and-a-half- by four- by three-inch cross. What's the big deal? I got plen-

ty of them in stock." He laughed out loud and set the oddball fitting back on the floor.

"You want three-and-a-half-inch fittings? I got 'em." He raced
over to a bin and pulled out two elbows. "You want seven-inch
steel pipe? I got a length upstairs." He pointed. "You name it, I've
got it!" He laughed out loud. Bob and I looked at each other.
"We've got everything!"

I wandered over to one of the bins and saw a few dozen
Webster Sylphon thermostatic radiator traps from around 1910.
They were brand new. I remembered a few difficult steam jobs
where I would have killed for these. There were bins filled to the
brim with 1½" angle, steam-radiator valves. Each must have
weighed ten pounds.

"They don't make them like they used to, do they?" Bob mut-
tered.

"They sure don't," I admitted, hefting one of the brand-new
valves in my hand.

"Hey, look at this," Doug called from the other side of the
room. We walked over. "This is a Robbins Cantwell pipe
machine. Ain't it beautiful? It's belt-driven, and all these gears
are made of wood." He poked the gears with his beefy finger.

"Does it still work?" I asked.

"Sure does! I use it all the time," he said. "That's the magic of
this place. I can find just about *anything* I need to get a job done
down here. Other guys go crazy when they have to work on an
old steam or hot-water system. Not me! I've got everything I
need right here. Never a problem!"

Doug went on to explain how his ancestors had catered to the
many paper mills that once upon a time called Turners Falls,
Massachusetts home. The mills all used steam, lots of steam, and
to them, time was always of the essence. The company that had
the pipe, valves, fittings and specialties in stock got the work.
The others didn't.

Over the years, George Starbuck & Sons accumulated stuff—
lots of stuff. The mills left, but the stuff stayed. Today, George
Starbuck & Sons is a regular heating museum. Everything I've
ever read about or heard about or talked about is there. And
much of it is brand-new.

"You ever see anything like this?" Doug asked as he stopped at
a porcelain commode with a wooden seat. He lifted the china pot
out of the commode by its steel handle. "This is how you flushed
it," he said with a chuckle. "And how about that!" He pointed to
an old Quiet May oil burner from the 1930s sitting in the corner.
"Or this!" He lifted up an old soil-pipe cutter. "I got the whole set

of tools for doing that type of work," he said, getting Bob's full attention.

"I remember when we used to work only with lead," Bob reflected.

"So do I," Doug said. "I remember this one time when I was younger. I was doing the plumbing on a job, and there was this old heating man working there too. I'd have the lead cooking in my pot and he'd come over and stick his finger right in the lead! 'Not hot enough yet,' he'd say and walk away laughing. I couldn't for the life of me figure out how he did it. That lead was over six-hundred degrees hot. Then one day I got it. I saw him sucking his finger before he put it in the pot. That's how the trick works. When your finger's wet, it keeps the lead from sticking. If you do it real fast, you don't get burned."

"Did you ever try it?" I asked.

"Oh yeah! Right away, I tried it. I wet my finger and stuck it deep in the pot. There was just one problem, though."

"What's that?"

"I forgot to wet under my fingernail." Doug let loose an enormous laugh as he remembered the pain. "Boy, you shoulda seen me dance! That hot lead got under that nail and chewed right into the finger. Whoa!" He shook his index finger, remembering the day.

"I guess that's how we learn," Bob offered.

"You're right," Doug admitted, "but that's the trouble with a lot of the guys in the business nowadays. They don't want to learn and they don't want to stop and think. All they want to do is replace parts. And you know why? Most of 'em don't know what the heck they're looking at. That's where they're missing out. If you have it in your head...you own it." He tapped his forehead lightly. How true that is, I thought.

We wandered over to some old Detroit regulators, controls that once used the steam from the boiler to operate the gas valve. "Ever seen these?" Doug asked.

"Only in books," I admitted.

He had some of those beautiful old 12" brass pressure gauges you see on jobs and try to snatch so you can take them home, clean them up and hang them on your wall. You know the kind I mean?

Doug Starbuck had them in stock, brand new.

"One time I went out to a house they were building on the other side of town," he said. "The carpenter was working with this kid apprentice. I got there at about three o'clock in the afternoon and sat down on a nail keg. I sat there for about an hour. The apprentice kept looking over at me. Finally, he asked the carpenter what I was doing. The carpenter looked over at me and went back to hammering. 'Doug's working,' he told the kid. 'Leave him alone.'

"The next morning I showed up with my drill and punched out all the holes I needed to do the whole job. I put the drill away and never took it out again. When I was piping the job, the carpenter said to the kid, 'See? You work first in your head, and then you pick up your tools. That's the *right* way to do things.'" If you have it in your head...you own it.

George Starbuck, Sr. had two sons, R.M. and George, Jr. R.M. never worked in the trade, he just moved down to Hartford and became famous as a book writer. He used what his father had taught him.

George Jr. stayed, worked the business for years and had a son of his own. He called him Joe.

Joe begat Lloyd and Lloyd begat three sons, Doug being one of them. Doug stayed; the others moved on into other fields. Doug had kids but the world called them to do other things. So Doug's it, the last Starbuck.

Somewhere along the line, one of the Starbucks installed this monster of an American Radiator Company hot water boiler. The boiler serves a gravity-fed Honeywell System that includes Honeywell's famous, mercury-filled Heat Generator, a device that, I'll bet, no more than a handful of the heating professionals in America know about. The Heat Generator was Honeywell's first significant invention, the predecessor to the hot water circulator.

Doug has one on his boiler. Pretty cool, eh?

They put a "fuel saver" coil in the breeching a long time ago to transfer some of the heat from the flue gasses into the system water. Naturally, it made the gasses condense and it's long since rotted out, but it's still hanging there as a good lesson in basic science.

The pipes are large and old and covered with asbestos. The heated water rises up by gravity with great efficiency to warm the rooms on both cold and mild days without firing the boiler too often. There's an open tank in the attic with an overflow pipe that sticks out through the roof. It's all still there, every bit of it, exactly as it was installed back in, oh, I'd say 1920. It's *all* there.

You want to go someplace to learn about the heating business, where we've been and what it all means? You want to find out what it is you're looking at when you wander through those old basements in your own town? You want to see what the Dead

Men were capable of doing? Then drive up Interstate 91 and take the exit that leads to the Connecticut River. Cross the bridge and knock on Doug Starbuck's door. You won't be sorry.

Bob and I sat for a while and talked with Doug's parents who are in the eighties and have minds as clear as mountain air. They told us stories and recollected dates and events, things that took place long before we were born. They reminded me, once again, why I do what I do. Reminded me that it's about more than just money—a lot more.

Before we left, Doug took us to the cemetery and showed us the graves. They're all there, all the Starbucks who came and went and left a bit of themselves behind in every job they did. It was a beautiful fall day and the yellow leaves rested quietly on the gravestones.

Bob and I looked at each other. If we stood real still, we could almost hear the Dead Men speak.

Listen.

INDEX

Sources

Beyond this list of books there are the countless conversations with people who know so much more than I do. I hope I've done a reasonably good job of retelling their stories.

Adlam, Napier T., *Heating by Radiant Means*. New York: Sarco Company, Inc., 1938.

Adlam, Napier T., *Radiant Heating*. New York: The Industrial Press, 1947.

American Radiator Company, *The Ideal Fitter*. New York: The American Radiator Company, 1925.

American Society of Heating and Ventilating Engineers, *The American Society of Heating and Ventilating Engineers Guide*. Baltimore, MD: The Horn-Shafer Company, 1926.

American Society of Heating and Ventilating Engineers, *The American Society of Heating and Ventilating Engineers Guide*. Baltimore, MD: The Horn-Shafer Company, 1934.

American Society of Heating and Ventilating Engineers, *The American Society of Heating and Ventilating Engineers Guide*. Baltimore, MD: The Horn-Shafer Company, 1937.

American Society of Heating and Ventilating Engineers, *The American Society of Heating, Ventilating, Air Conditioning Guide 1948*. Baltimore, MD: The Horn-Shafer Company, 1948

American Technical Society, *Cyclopedia of Heating, Plumbing and Sanitation*. Chicago, IL: American School of Correspondence, 1909.

Allen, John R. and Walker, J.H., *Heating and Ventilation*. New York: McGraw-Hill Book Company, Inc., 1922.

Bell & Gossett Company, *Handbook*. Morton Grove, IL: Bell & Gossett Company, 1949

Bernan, Walter, *On the History and Art of Warming and Ventilating Rooms and Buildings, Volumes 1 and 2*. London, England, 1845.

Carpenter, Rolla C., *Heating and Ventilating Buildings*. New York: John Wiley & Sons, 1895.

Chase Brass and Copper Company, *Chase Radiant Heating*

Manual. Waterbury, CT: Chase Brass and Copper Co., Inc., 1945

Daniels, Ara Marcus, *Steam and Hot Water Heating.* Washington, D.C: Domestic Engineering Publications, 1928.

Donaldson, Wm., *Modern Hot Water Heating, Steam and Gas Fitting.* Chicago, IL: Frederick J. Drake & Co., 1906

Fuller, Charles A., *Designing Heating and Ventilating Systems.* New York: U.P.C. Book Company, Inc, 1923.

Giesecke, F. E., *The Design of Gravity-Circulation Water Heating Systems.* Austin, TX: Heating and Ventilating Magazine Company, 1926.

Graham, Frank D. and Emery, Thomas, J. *Audels Plumbers and Steam Fitters Guide #3.* New York: Theo. Audel & Co, 1925.

Harding, Louis Allen, and Willard, Arthur Cutts. *Mechanical Equipment of Buildings.* New York: John Wiley & Sons. Inc., 1929.

Honeywell Heating Specialty Co., *How to Properly Design & Install the Honeywell Method of Hot Water Heating.* Wabash, IN: Honeywell Heating Specialty Co., 1912.

International Textbook Company, *I.C.S. Reference Library.* New York: Burr Printing House, 1905.

ITT Bell & Gossett Company, *Hydronic Heating and Cooling Systems.* Morton Grove, IL: ITT Bell & Gossett Company, 1964.

Jehle, Ferdinand, *Hoffman Data Book.* Indianapolis, IN: Hoffman Specialty Company, 1951.

Johnson, John W., *Johnson's New Handy Manual.* Chicago, IL: Frederick J. Drake & Co., 1928.

King, Alfred G., *500 Plain Answers to Direct Questions on Steam, Hot Water, Vapor and Vacuum Heating.* New York: The Norman W. Henley Publishing Company, 1923.

King, Alfred G., *Practical Steam, Hot Water and Vapor Heating and Ventilation.* New York: The Normal W. Henley Publishing Company, 1912.

Leeds, Lewis W. *Ventilation.* New York: John Wiley & Son, Publishers, 1868.

Lenman, Tomas, *Water and Pipes.* Virsbo, Sweden: Wirsbo

Bruks AB

Reid, David Boswell, M.D., *Illustrations of the Theory and Practice of Ventilation.* London, England, 1844.

Starbuck, R.M., *Questions and Answers on the Practice and Theory of Steam and Hot Water Heating.* Hartford, CT: The Bond Press, Inc., 1905.

Tidd, Ed, *Tidd-Bits, Factual Reports on Hydronic Trouble Installations with Tips for Their Prevention.* Morton Grove, IL: Bell & Gossett Company, 1960s.

Check out our other books
for regular human beings.
You'll find us on the Internet at
www.HeatingHelp.com
Or call for our catalog.
1-800-853-8882.